MW00778342

Birds, Sex and Beauty

Birds, Sex and Beauty

The Extraordinary Implications of
Charles Darwin's Strangest Idea

MATT RIDLEY

HARPER

An Imprint of HarperCollins*Publishers*

HarperCollins books may be purchased for educational, business, or sales promotional use. For information, please email the Special Markets Department at SPsales@harpercollins.com.

Originally published in Great Britain in 2025 by 4th Estate, an imprint of HarperCollins Publishers.

All photographs are by the author except where stated otherwise.

FIRST U.S. EDITION

Library of Congress Cataloging-in-Publication Data has been applied for.

ISBN 978-0-06-334298-9

25 26 27 28 29 LBC 5 4 3 2 1

To the memory of Wonky Tail

Have you heard the blackcock's husky crow
In the cool grey light of morning,
When the mists were on the vale below,
And the mountaintops were all aglow
With ruddy gleams that seem to show
The pathway of the dawning?

Have you stopped among the heath and ling
To see the greyhen stealing
With her speckled poults of tender wing
That closely to the covert cling
And fear to take the final spring,
Their whereabouts revealing?

ANON

Contents

Preface

The sea lifted smooth blue muscles of wave as it stirred in the dawn-light, and the foam of our wake spread gently behind us like a white peacock's tail, glinting with bubbles.

GERALD DURRELL, *My Family and Other Animals*

Intelligent life on a planet comes of age when it first works out the reason for its own existence.

RICHARD DAWKINS, *The Selfish Gene*

Two books that left a big mark on me were *My Family and Other Animals* by Gerald Durrell and *The Selfish Gene* by Richard Dawkins. One helped turn me into a naturalist as a child, the other into a scientist as a student. Both authors are superb writers: exquisite manipulators of syntax, vocabulary and metaphor. I was lucky to get to know them both at various stages of my life. But whereas Durrell was a witty witness to the beauty of nature, infecting the reader through a form of magical realism with a fascination for wildlife of all kinds, Dawkins is a penetrating analyst of underlying truth, persuading the reader to understand the surprising implications of evolution in fresh ways.

I have long been frustrated by the wall that divides scientific analysis from natural history writing. For me nature is both magical and fascinating, both beautiful and true. To understand it you must dive deep and dissect the genes, the molecules and the numbers; but to appreciate it you must stand back to watch and experience nature in awe and wonder. To do one without the other is, for me, deeply unsatisfactory. The two need not be in conflict: to borrow from the title of one of Dawkins's books, to unweave the rainbow by understanding how it comes to be is to make it still more marvellous, whatever John Keats said to the contrary.

I decided to write a different book from all my previous ones about science: to attempt the kind of natural history prose I have loved since childhood, while still delving into the science behind it. *Birds, Sex and Beauty* aims at an ambitious fusion of the genres of natural history writing and science writing. I have deliberately interspersed the two styles. Chapter by chapter, I chronicle a single spring morning at a single spot on a lonely hill, watching a single group of extraordinary birds. But alongside this, I use what I see to open and explore a trove of insight, painstakingly acquired by diligent scientists over many years. In the process I have also attempted a history of the vexed theory of sexual selection by female choice, laying bare a century of disagreement about an idea so powerful, so weird, so wonderful, that I still feel unsure whether I have gone far enough in accepting all its implications.

As the book progresses I bring the two paths together and, taking wing, in the last chapters become a bit of a travel writer, flitting from the Arctic to the equator in search of more birds and more insights. All in the hope of persuading you, reader, that mate choice is the force that has shaped some of the most beautiful and bizarre features of the world – including perhaps the very mind you are using to comprehend these words.

1.

The Lek

It was the lark, the herald of the morn,
No nightingale. Look, love, what envious streaks
Do lace the severing clouds in yonder east.
Night's candles are burnt out, and jocund day
Stands tiptoe on the misty mountain tops

WILLIAM SHAKESPEARE, *Romeo and Juliet*

4.30 a.m., April, the Pennine hills

Saturn is just bright enough above the western horizon to penetrate a thin veil of cloud. The moon has set but a faint glow stains the eastern skyline and my eyes have adjusted to the gloom. Under foot the grass is crisp with a glitter of frost. I walk fast, eager to reach my destination before the light does. The rushes are uneven and I stumble a few times. It's too dark to make out the path but I know the way. The pregnant, reverent silence of the hour before the dawn, when even the wind drops, is broken only by my breathing. It's too early for the larks. Then an urgent, unearthly humming above me in the sky breaks the stillness: a reverberation straight from science fiction, like the engine of a descending spacecraft. Or, in reality, a drumming snipe, the sound coming not from its beak but from stiff outer tail feathers vibrating as the bird dives downwards from a height.

The little canvas hide looms through the darkness. My gloves fumble with the zip but I soon manage to crawl inside to sit on a small camp chair. I open the front flap, tie it up and listen, holding my breath. There is no sound; the snipe has gone. Using a thermal imager I scan the grass in front. A distant sheep glows bright but otherwise the eyepiece shows the vegetation as a mess of blurry dark greys against a still darker, colder sky. A few minutes pass and I scan again as the first Skylark begins to sing and a distant Curlew serenades the coming dawn. This time a bright spot is visible a short distance away, moving slowly towards me. As the spot emerges from the grass it resolves itself into the shape of a bird, its head bright with heat, its body lukebright.

Suddenly two other birds appear in the imager's viewfinder, flying towards the spot where the first bird is standing. A sharp sound 'pierces the night's dull ear', as Shakespeare once put it, a strange cry, again of exactly the kind a film director might give to an angry alien. It came from the first bird. It's neither a sneeze nor a hiss, neither a squawk nor a screech, but somewhere in between all four: 'Tschu-wee'. One of the other birds answers with the same call and soon there is a cacophony of calls as more and more birds join. Then, more softly but continuously, a new sound begins: 'Hoo-doodle-woo-doodle-WOO-hoo'. It is a still more otherworldly call, somewhere between humming and crooning, bubbling and cooing. As if the aliens have started the hyperdrive on their spacecraft and the engine is idling. Or maybe it is more like an incantation chanted in a distant chapel, which in a way it is. It's the mating call: 'Please God, bring me a mating opportunity today!'

I can now just make out the birds with the naked eye, their black bodies and white bums blurrily visible to the rods in my retina, if not yet to the cones in my fovea, against the winter-bleached grass in the monochrome twilight. They are male Black Grouse, chunky birds the

Somewhere between humming and crooning, bubbling and cooing.

size of large chickens. Later, when the light grows brighter, I will count fourteen cocks in a space no larger than a tennis court, plus three young ones that turn up late and try to gate-crash the party. They are attracted to this exact spot before first light every morning for nearly eight months of the year, from October to June. This strange dawn ritual, as thrilling as any druid ceremony – no, far more so! – happens right here every year.

I am here to eavesdrop on a bizarre natural phenomenon, a 'lek'. The word is Swedish, meaning 'play'. An old English word 'lacan' also means to frolic and one Victorian writer, Henry Seebohm, called the Black Grouse arenas 'laking places', using a Yorkshire-dialect version of the word that is still sometimes used. These days, in English at least, the word 'lek' has displaced its various rivals such as the German 'balz' or 'spel' and is used for communal sexual displays in any species of animal anywhere.

The cocks gather in a space no larger than a tennis court.

What I am watching may resemble play but it is deadly serious. It's a sex market, in which the male suppliers of sperm fight for the right to satisfy very discerning female customers and usually only one male succeeds. Every day between March and June (and less frenetically between October and February) these same birds will gather at this one secret spot on a grassy moor in the Pennine hills and take part in a frenzy of dancing, singing and occasionally vicious fighting for about three hours in the early morning. Each bird – unless or until he is injured, or promoted when his neighbour is wounded or killed – will occupy the exact same square on the chessboard all season, furiously defending his little patch and displaying incessantly for the benefit of any visiting female, a slave to his horny hormones. His home territory on the lek is just five to ten paces across. The more senior he is, the more central but smaller his patch, and the more continuously he displays.

A lek is like nothing else in nature. Most wildlife is constantly on the move even within a limited home range but these birds or their ancestors have been here at this very same place day after day and year after year perhaps for decades. Most birds in the breeding season scatter into territorial pairs, but Black Grouse males gather in this tight but tense flock. Most creatures indulge in brief and occasional displays at the start of the breeding season, but these ones dance and sing for hours with hardly a break and do so for many weeks on end. In most birds the act of copulation is almost impossible to witness – should you be interested in such things – but at a lek the date is fairly predictable to within two weeks, the time very predictable (it happens within twenty minutes either side of sunrise), and you can see several instances of coition in the same location on the same day.

Even better, the performance is concentrated around that magical, mysterious moment of dawn, when the half-dark, the sounds of springtime and the absence of anything human seem to transport you back in time. In this moment, 'when creeping murmur and the poring dark fills the wide vessel of the universe', as the bard put it, I feel I could be back in a primeval age when this Pennine hillside was yet to be trodden by Neolithic or Iron Age tribes and their cattle or sheep.

About four million times since the great ice melted on these slopes for the last time, the sun has dimmed the planets over this landscape. On tens, perhaps hundreds, of thousands of those dawns there was probably a gathering of Black Grouse at this very spot, so wedded are the birds to their traditional sites. Yet on almost all of those occasions the place has been empty of people. Perhaps stone-tooled hunter-gatherers lay in umbered ambush by this very lek in the warmer springs of nine thousand years ago, at the time of the Holocene Climatic Optimum, their yew bows and flint-tipped, hazel arrows ready to kill a bird for food, or to use its feathers in ceremonial outfits. Perhaps a few thousand years later druids stood here on this hill, inferring spiritual meaning in the shadows and chanting the secret whispers of their fertility prayers just as the Black Grouse are now doing, but even they probably did so at the solstice, not around the equinox, and probably on a different hill. Perhaps shepherds camped here in medieval times, while herding sheep to market in the lowlands to the east and watched by the paly flames of their fires as the sky faded from deep indigo through pale purple to cool lemon and then fiery orange at the eastern horizon. That pure colours are beautiful is something human beings and birds can agree on, or so I shall argue in this book. Perhaps that terrible cad, Andrew Stoney Bowes MP, the model for Thackeray's Barry Lyndon, galloped past here after kidnapping his abused and escaped wife, the impossibly wealthy heiress Mary Eleanor Bowes, ancestress of the current king, in the bitter cold of February 1785: for this land was once part of her inheritance and he was eventually arrested near here by one of her loyal tenants. Sexual coercion is rarer in Black Grouse than in people. Surely somebody at some point listened to the sneezing and bubbling sounds in the dim dark and wondered what they were all about. As I do now.

The four o'clock alarm was painful, but it means I have the world to myself. I feel a twinge of guilt at ogling another species's ancient fertility rite: both the Black Grouse and I are here because (their) sexual intercourse happens at first light and on just a few days a year. I'm a voyeur, even if not for salacious reasons. (There's nothing remotely erotic about wearing thermal underwear, a woolly hat and boots while crouching in a little camouflaged tent at such an early hour.)

It's still pretty dark and the eerie chorus of bubbling (known as 'roo-kooing') is punctuated by the Tschu-wee sneeze call. Old naturalists used to describe this sound as like the whetting of a scythe, but my friend Tarquin came up with a much closer modern simile: it's like opening a can of lager, only louder. The birds emit the noise while stretching tall and opening their wings, whereas the roo-kooing comes when they bow close to the ground, their necks swollen with gulps of air pushed into special pockets in the lining of the throat. Sometimes,

if they think a female is approaching – even if it turns out to be a passing Curlew instead – they leap into the air with a sort of flutter-jump,
accompanied by a vibrato version of the lager-can sneeze. A little later
in the morning, as the light looms, the flutter-jumps will grow in
frequency. Now comes yet another sound, harsher and plainly irritable:
'naair-na-naaair' or 'choc-ke-ra-da', as it was rendered by one observer
a century ago (neither rendition does it justice). It comes from two
males squaring up for a fight. Content warning: there is violence as
well as sex in this play.

The lek may be in the exact same place every year but there seems
to be nothing special about the location. The vegetation, the eleva-
tion, the gradient and the view are all unremarkable: near the top of
a hill but not on top, level ground but not dead flat, good view in all

They leap into the air with a sort of flutter-jump.

directions, but not especially far. The ground vegetation is mossy but with patches of grass and rush and heath, some of which are trampled bare. Scattered around are the seeds of hawthorn, left here in the autumn in Black Grouse droppings. In Finland leks are often on frozen lakes.

I am here to watch in wonder, but also to wonder why. Why here, why this species, why not others, just plain why? There is a theory, named sexual selection, that has the power to explain some, perhaps all, of what I am seeing and not just in this one bird, but in the whole of aesthetics, even in the human world. But the theory's nooks and crannies are still controversial, uncertain and mysterious. Science is always at its most thrilling when unsettled: it is the process of tackling mysteries, not the habit of accumulating facts. And it is no exaggeration to say that this enigma is one of the great unanswered questions of evolutionary biology. Perhaps it seems a bit hyperbolic of me to pretend that the dancing grouse that I see before my hide are on a par with, say, grand unified theories of physics, the source of consciousness, or the cause of ice ages – but hear me out. The existence of beauty, by which I mean the fact that most human beings can agree on some things being exquisitely pleasing to the eye or ear in a universal and timeless way, is a great scientific mystery in itself and one that no philosopher has satisfactorily illuminated. Yes, there are ideas but none of them really gets to the root of the matter, not yet.

The fact that we human beings also find some animals especially beautiful adds to the puzzle, and the fact that the most striking examples of this cross-species aesthetic appeal are animals in the throes of erotically arousing their conspecifics makes it still more intriguing. Beauty seems to matter to birds too: nobody can watch a Peacock displaying to a Peahen and not be struck by the apparent role that beauty is playing. But there is not even the beginning of a satisfactory explanation yet for the fact that the bird I am watching is black and white, rather than say red and yellow, or that it has bright red eyebrows, rather than say a blue bib, or that its tail is shaped like a lyre, rather than straight or forked.

Think about that for a minute. Astronomers reckon to explain why a star or a planet has the particular character that it has, using general principles derived from theory. Physicists can explain why particular elements have particular properties, chemists why particular compounds undergo particular reactions, biologists why particular genes are recipes for particular proteins. General theories allow the illumination of specific examples. But here is a full and venerable theory of evolution by selection that cannot begin to explain the particular eccentricities of the feathers and the frolics of a Black Grouse, or a Peacock or a singing Nightingale. The protagonists in this debate cannot even agree on the main mechanism at work. Not for want of

Beauty seems to matter to birds.

trying. Five generations of biologists have now wrestled with this problem, some of them with observations, some with experiments, some with mathematics. A few have confidently declared the matter solved, but always prematurely. Yes, this is a great scientific question whose interest extends far beyond the bird-watching fraternity. By the end of this book I hope to have persuaded you that this theory holds the key to understanding the human mind. But this is not a book about one self-obsessed, African ape.

'The lek paradox'

In all animals mating is a deal: one sex donates a few million sperm, the other a handful of eggs, the merger between which – unless a predator intervenes – will result in a brood of young. Win-win for the parents, genetically speaking. But there are few creatures that behave as if sex is a dull, simple or even mutually beneficial transaction and many that behave as if it is an event of transcendent emotional and aesthetic salience to be treated with reverence, suspicion, angst and quite a bit of violence. In the case of Black Grouse the males dance and sing for hours every day for several exhausting months, selling their little packages of sperm as passionately and persuasively (and frequently) as they can. To prepare for the ordeal they grow, preen and display fancy, twisted, bold-coloured feathers. They gather together in one spot, putting themselves at conspicuous risk of attack by hawks and forgoing opportunities to feed. They fight with deadly intent again and again, suffering significant injuries. As excitement builds they expand the bright red, swollen, fleshy combs over their eyes, covered with hundreds of tiny tentacles like vermillion sea anemones. The act of sexual congress itself, the consummation of the deal, takes seconds. The rest has taken months of practice and preparation and is elaborate, extravagant, exhausting and elegant. Why?

The full answer to this question is, as I say, still mysterious. Evolutionary biologists can explain why males are generally (not always) the eager sellers, females the discriminating buyers of gametes. They can begin to explain why some species have gaudy males, others beefy ones, still others dull ones. But they struggle to explain – or at least they disagree passionately – why it is that in some species this extravagance goes beyond the mere gaudy and into baroque and bizarre shapes and postures. And as I say they don't seem to be able to satisfactorily explain why bird displays generally seem beautiful to us.

Among the enduring enigmas, Black Grouse represent the purest example of one. For here is a bird that is obsessively choosy – on the females' part – when it comes to sex, and in which a single male gets almost all the matings at any one lek. Given that this also happened a generation before, and that males live near where they were born, it follows that some of the rivals on this lek will probably be brothers and half-brothers, fathered by the same male. For generation after generation the local population goes through a narrow genetic filter, with just one male doing most of the fathering each year, so the genetic diversity in the population is inevitably low. There must be relatively little to choose between the males genetically compared with other species. So in this species, far more than others, there is surely little point in being choosy.

Yet female Black Grouse are – as I say – obsessively choosy, breeding only from the best, as if the males were racehorses. Indeed, racehorses suffer from the same paradox, because breeders pay gigantic sums for access to the sperm from the best stallions despite the entire breed being massively inbred ('thoroughbred') from a handful of original stallions with very little genetic variation compared with other breeds of horse.

This puzzle has been known to biologists since 1979 as the 'lek paradox' and it remains unresolved. The scientist who coined the term, Gerald Borgia, was drawing attention to the fact that if females are consistently choosing the same males, they will soon cause any variety to diminish, maybe even disappear. So the choosiest species seem to have the least reason to be choosy. Resolving this puzzle to my own satisfaction is one of the tasks I have set myself by watching this lek. Incidentally, unlike in racehorses, lekking does not make the species problematically inbred, because female Black Grouse emigrate from their native area so they bring fresh blood to each lek. The females attending this lek were probably fathered in a different dale, twenty miles away or so. But a glance across the lek now confirms that there is very little to choose between the males and unless you carefully count missing tail feathers or notice scars about the mouth, you cannot even recognise individual males.

As I watch, I ponder a human analogy for the lek paradox, though very inexact because human beings are not a lekking species. Imagine an unmarried young woman, Miss Bennet, turning up in a village in which there are fourteen single men, many of whom are half-brothers or cousins, all of whom look roughly alike, wear identical frock coats and live in similar small houses on the main street: Tom Darcy, Dick Darcy, Harry Darcy, Fitzwilliam Darcy, etc. ... There is nothing to choose between them in wealth or breeding, and very little in looks or

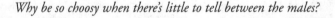

Why be so choosy when there's little to tell between the males?

dress. They are all qualified chartered accountants. Though they spar continuously among themselves, the Darcys are scrupulously polite and attentive to the new woman, each one bursting into the same rendition of 'O Sole Mio' to serenade her whenever she passes down the village street. Yet somehow every woman who has arrived in the village before has agreed that Tom Darcy is the man for her, and the rest are nothing. The village is full of young ladies bringing up Tom Darcy's babies. Have they decided this independently or are they copying each other? We don't know. Our protagonist Miss Bennet, on learning this, obediently follows suit and submits to the same gentleman's caresses. Why?

I am here watching the lek for the umpteenth time mostly because I love being here – watching rare birds at close quarters, while being serenaded by an avian dawn chorus may be as close as I get to feeling spiritually uplifted, strange fellow that I am. But also I am convinced that one day, just by watching very closely and paying attention to the details, I will finally resolve not only the lek paradox but the full mystery of sexual display in birds, and therefore of the origin of beauty itself, to my own satisfaction at least. To do this I really should have captured each male bird, then weighed, measured, marked and genetic-fingerprinted him, and staked out the lek with evenly spaced markers. I did a good bit of exactly this with other species of bird many years ago when I was a professional scientist but today I am being more of a naturalist.

Fortunately, a former Oxford colleague of mine by the name of Rauno Alatalo began to do exactly these scientific things at Black Grouse leks in his native Finland, and at a separate site in Sweden, more than twenty years ago. Although Rauno suffered a stroke in 2008 and sadly died in 2012, his legacy continues as colleagues are still trying to understand this enigmatic species. The Nordic scientists catch Black Grouse in traps baited with oats during the late winter and spring in an area of bogs, lakes, forests and open areas used for peat digging.

For each male, they weigh the bird, estimate its age from its plumage and measure the length of its wing and the length of its lyre-shaped tail.

Samples of feathers are cut from the breast and in 2006, Heli Siitari took a few sample feathers from each bird back to a laboratory, placed them on a black velvet background in a dark room and illuminated them with a xenon fibre-optic light source at a consistent angle so that a spectroradiometer could read out the blueness of the feather: the peak reflectance is in the violet-blue part of the spectrum. (Some bird species can see in the ultraviolet, but in grouse the peak of their short-wave vision is only a little shorter than that of human beings.) Later work has also estimated the size of the fleshy red combs above the eyes and measured just how red the combs are, using a spectrophotometer. For some birds the scientists also extract a sample of blood for genetic and hormonal analysis and to measure parasites. Females are also weighed and blood-sampled. The birds are then released with a metal ring and a couple of coloured plastic rings on their legs for recognition. Over the years the Finnish study has trapped and measured more than 1,200 females and 1,700 males, many of them more than once.

Meanwhile the diligent, and presumably sleep-deprived, scientists literally stake out five or more leks, with white markers placed to create a grid on each one. Every day for two weeks during late April and early May from 3 a.m. to 8 a.m. they watch and video the leks, recording each male's position and behaviour as completely as possible: its distance from the centre of the lek, its total time at the lek each day, how often it fights and displays, and of course whether and when it mates. The colour rings allow individuals to be identified at distance and any matings and behaviours assigned to a specific individual. This exhaustive, invaluable study, to which I will refer repeatedly through-out this book, has enabled somebody like me to interpret in far more detail what I see on the lek I watch, where I can only identify some of the individual birds from a few distinguishing marks. The information it has revealed is remarkable and far from taking the mystery out of nature, it only makes it more magical, in my eyes at least.

'Do not long retain the attention of their sable lords'

Back at my lek, a clockwork cackle announces the arrival nearby of a cock Red Grouse. He is smaller and far less flamboyant in appearance than his Black Grouse cousins. Their last common ancestor lived about six million years ago, roughly the same time as our last common ancestor with Chimpanzees. True, the Red Grouse also sports a pair of bright orange-red eyebrow combs at this time of year and he is noticeably more vocal than his mate but otherwise he is hardly different from his spouse in size or appearance. His cocksure call, delivered while stalling in the air about twenty feet off the ground after a short flight, announces that he owns a territory here. It will be much larger, an acre at least, than that of a Black Grouse – encompassing all fourteen Black Grouse territories on this lek and some more land besides – but unlike the Black Grouse he will largely stay within its boundaries all day. He will have defended it all winter – except when it was covered in deep snow – for the exclusive use of a single hen, who is already incubating eggs at this date, a month earlier than Black Grouse.

The Red Grouse does not lek: he does not gather in the same place as his rivals and his courtship displays are simple and short; his plumage is workaday practical camouflage, designed to blend with heather, with only a hint of burnt-copper reddish-brown masculine smartness. (Males are much more variable and easy to identify as individuals than Black Grouse, by the way: some have more white on the belly or chin, some are darker, others more mottled; they have not been 'thoroughbred'.) The male Red Grouse defends this large territory full of food to support a single mate; he then stays with the brood when the eggs hatch, keeping watch as the chicks grow. Sex is therefore a much more utilitarian transaction: I'll protect the food supply, he implicitly says to the female, and help look after the kids if you let me father them.

The female Black Grouse, by contrast, will get nothing from her mate either before or after the act of sex. As William Yarrell put it in 1843, female Black Grouse 'do not long retain the attention of their sable lords'. Telling me: the pair bond, which lasts six months in Red Grouse, lasts less than six seconds in Black Grouse. The male defends and provides no food and supplies no parental care. The female is a single parent, obsessed, it seems, with finding as a mate a male who displays well near the centre of a lek but is otherwise no use to her. Why are the two birds so similar in their appearance and ecology yet so different in their mating habits? What is in it for her to choose such a dandy and cad? In our hypothetical village, Tom Darcy has seduced a dozen women already but never once looked after any of their children. And the women don't seem to mind.

Does she actually choose, or do the males choose for her? Nineteenth-century naturalists routinely described the Black Grouse lek in terms of medieval jousting. The cocks fight at the 'lists' or 'tournaments' till the best knight wins and claims the damsel's hand. The female's role is passive – and certainly untainted by anything as vulgar as lust or as active as choice. Here's the Newcastle engraver Thomas Bewick, writing about Black Grouse in 1805, for example:

> The males assemble at their accustomed resorts, on the tops of
> high and heathy mountains, when the contest for superiority
> commences, and continues with great bitterness till the
> vanquished are put to flight; the victors, being left in possession
> of the field, place themselves on an eminence, clap their wings,
> and with loud cries give notice to the females, who immediately
> resort to the spot.

This is wrong in most particulars. Apparently, it took some leap of the imagination for nineteenth-century chauvinists to see females choosing males based on beauty, when it was clear that in their own society men were choosing women based on beauty. Even Alfred Russel Wallace, the co-discoverer of natural selection, who was known as a champion of the rights of women, struggled with this. So they tended to fall back on the idea of the males deciding among themselves who would be 'chosen' by the female instead. In the 1880s, Henry Seebohm, a Yorkshire quaker and steel maker who travelled to Siberia in search of rare birds, described Black Grouse leks and quoted a Mr Dixon as saying that as the morning progresses, the 'females are getting more interested every moment, ready to bestow their affections on the victorious males ... And so the combats proceed until all the brides are won, when the strife ceases.'

Ready to bestow their affections on the victorious males.

The artist John Guille Millais in his *Game Birds and Shooting Sketches* (1894) described rousing himself to light a candle and 'dress in the cold grey morning' in order to visit a lek more than once. He saw the whole thing through an Arthurian prism. He was transported straight to Camelot as he watched a male commencing 'his song of war, at which some wandering knight, who has yet his laurels to win, soon takes offence and at once challenges him'. When the female arrives, 'the male that is nearest to her pairs with her and fights off any other that disputes his possession. The hen meanwhile walks sedately round her lord and master, picking about at the grass coquettishly.' Wrong in almost every detail, I am sorry to say.

In 1924 the Northumbrian naturalist Abel Chapman described the 'tournament' of Black Grouse as a 'mystery oft described but never yet explained'. He thought the lek could not be about sex, because 'despite all this conspicuous passion-play, watch it as often and as long as you will, yet neither the author nor dozens of gamekeepers and hill-shepherds (to whom these are everyday spectacles) ever see a sign of contact between the principal actors – and in broad daylight it could not occur unseen'. Well, I have seen plenty of contacts between males and females.

The Arthurian myth persists in some quarters to this day. When the BBC aired a series of films about British wildlife in 2023 called *Wild Isles*, they included a sequence from a Black Grouse lek in Scotland. The commentary told us that we were witnessing a male with missing tail feathers, called Half Tail, newly arrived on the lek, challenging an incumbent for the top spot and winning, so that he then got to mate with the females. Not only did this revert to the old Victorian habit of implying that females had no say in the matter, it also badly misread what happens between males as a win-lose fight, which it almost never is. In the film I saw a normal fight between two neighbouring males on the lek with no change in their status and no evidence of a change in which bird would get to mate. Here is Sir David Attenborough's commentary, straight out of Malory's King Arthur: 'This particular lek in the Cairngorms is dominated by a formidable male nicknamed "the Boss". Each morning he fights off challengers. But now a new cock on the block has arrived, Half Tail … And now he sets his sights on the Boss's crown. For the past week the two of them have been sizing each other up and now it's the showdown [footage of a fight]. Half Tail has triumphed. This lek has a new boss.' No: the last shot shows two males simply disengaging from the fight as neighbours on a lek always do, with no displacement of one by the other. Storytelling in natural history documentaries is not new, of course, but does it have to be so Victorian?

As I say, these Arthurian accounts of Black Grouse leks are wrong in many details. The males do not fight first and court females second; both happen on and off all morning. The females do not submit to being a possession of the male; they choose their mate and invite copulation. There is little or no vanquishing or putting to flight. The males fight, yes, but virtually every fight ends in a draw with both combatants still occupying the same spaces when the females show up. Yearling cocks that try to join the lek by landing in the middle are seen off, but all the rest of the males just stay on their courts. The old idea that leks establish a hierarchy of dominance is misleading. As I can clearly see from my hide, each male is dominant within the invisible boundaries of his territory. No rival is allowed therein, but nor is he allowed within his neighbour's court.

So I am struck by the egalitarian nature of the lek, in contrast to the Arthurian legend. As one Finnish scientist put it in the early 1960s, the central males 'are equal to each other, which is shown especially by the fact that their fighting along territorial boundaries remains indecisive, and that they show outstanding respect for their common territorial boundaries'. Two Dutch scientists observed the same thing in the Netherlands around the same time: 'Each of the territorial males has free access to his own territory only, and is, in general, victorious over every trespasser. It can therefore be concluded that on the lek all territorial males are of equal rank; none has special privileges.' It is noticeable that when the males return to the lek after fleeing some brief alarm, they run through others' territories in apologetic and cowardly fashion before suddenly turning as they reach their own border, flaring their tail and puffing up their necks.

Some of the old naturalists were a little more careful than others in describing what they saw. Llewellyn Lloyd, a British sportsman who hunted a hundred bears in Sweden in the 1820s, wrote a book about the game birds and water fowl of Sweden and Norway, published in 1867, in which he said that the male Black Grouse gather at the 'lek-stalle' and fight, adding charmingly that the females watch these battles, uttering a 'dolorous cry, *ack, ack, aa,* expressive of their anxiety for an embrace'. It's true that females do make this call when flying to and from the lek. But Lloyd too thought the fighting ended in the loser fleeing the arena, which is not true, and he stated that mating itself happens away from the lek, which is also mostly untrue. Yet on the question of whether females choose, Lloyd was a little more perceptive than most: 'More than one ornithologist tells us that the hens never solicit the favours of the cocks; but the reverse is the fact.' He cites wild hen Black Grouse coming to visit a caged cock in the garden of a Mr Hultman, who watched from his bedroom window.

'The dampness of these English moors with their dreadful, stealing, chill mists'

Not till 1909 did a truly accurate and detailed account of a Black Grouse lek appear in which a naturalist wrote down what he saw, not what he thought he saw. It came from Edmund Selous, my favourite character in this corner of history, who deserves a far greater reputation in science than he got. I had barely heard his name till I began researching this book and started seeking out those who had written about

Black Grouse leks in the past. Selous's descriptions seem to have been comprehensively forgotten. He turns out to have observed and described pretty well every detail I noted in ten years of lek watching. I could have stayed in bed! And his unfashionable conclusions about the theory of sexual selection were right on the money.

The son of a wealthy stockbroker, this shy and reclusive Cambridge-educated barrister, who had early on decided that bird watching was more fun than lawyering, wrote long accounts of his many, many days observing birds alone for hours on end. His stamina was astounding, his powers of observation acute, his interpretations cautious, his scientific reading deep. His guiding belief was that naturalists did too much killing, skinning and measuring of birds and not enough watching them behave. His brother Frederick had no such scruples and was a far more famous big game hunter, after whom a huge national park in Tanzania is named to this day. Had Edmund joined the societies and clubs where scientists gathered, and had he not criticised colleagues as 'thanatologists' – or death studiers – he would quickly have gained a high reputation. But he hated company. He was 'odd, withdrawn, shy, and solitary in type and a man of marked idiosyncrasy', according to the ornithologist David Lack.

Yet he published numerous books, several of which were aimed at children in a unique prose style, his mother having been a talented and published poet. In his papers, lyrical and detailed descriptions of bird behaviour are interspersed with intimate accounts of the discomforts of bird watching. 'I have been without my plaids, indeed,' he writes after one cold morning watching a Black Grouse lek on an English moor, 'but warmly clad in an ordinary way, and with a motor suit overall – I may say here, en passant, as perhaps of use, that double trousers, shirts &c, are, in my experience, warmer than one, with underclothing; and the two methods are combinable.' He urged bird watchers to sleep fully clothed so they could spring into action early. (Perhaps surprisingly, he married and had three children. His great-grandson, Andrew

Selous MP, kindly tried to unearth family memories for me but without much success.)

In April 1907, Selous travelled to Sweden at the invitation of a distinguished politician, Niklas Biesert, to watch a Black Grouse lek. His host having been called away, Selous found himself in a large mansion by a lake in a pine forest with only 'obliging and accommodating' servants and a Svensk-Engelsk dictionary for company. With few human distractions, he came to live among the Black Grouse obsessively that spring. The forester showed him where to find the lek and the night watchman then woke him every morning at three or even earlier so he could make his way to a vantage point from which to watch for hours despite the bitter cold. No detail escaped his eye, and I find it fascinating to read him describing the very same patterns of behaviour that I have often seen more than a century later.

In early May 1908 the indefatigable Selous, now fifty-one years old, was back in England and spent ten days on a moor in Northumberland to watch another Black Grouse lek. After being shown by a gamekeeper to the location, with twenty males on it, seven miles from his lodging, he would start off on his bicycle at midnight or 1 a.m. to get there before the light, riding through wind and rain that he found – despite the motor suit and double layers of shirts – more trying than the dry Swedish cold; he lamented 'the dampness of these English moors with their dreadful, stealing, chill mists'. I know what he means. But his reward was rich. At this larger lek – there had been only half a dozen males at the Swedish lek he watched – he saw for the first time, just as I have seen, the jealousy between females when several are at the lek at the same time. They fanned their tails, like males did, and drove each other away from the courting male they preferred. He saw mating at last and noted how up to four rival males leaped upon a male when he mated, trying to interfere. I have often seen the same.

As I peer out of my hide I can see two cocks to the left squaring off for a confrontation at the mutual border of their territories, their heads held high, their necks stretched and thin, their calls irritable and high-pitched, 'choc-ke-ra-da', as Selous rendered it. They move rhythmically back and forth like boxers bouncing to keep clear of any thrust that may come. An all-out attack with the feet and claws and wings may be imminent, but more often there is a détente, and sure enough these two slowly turn away from each other with no actual contact and resume roo-kooing and sneezing. Six more cocks nearby are roo-kooing incessantly, their massively swollen, bright red eyebrow combs sharply contrasted with their black heads, their dark blue, tremulous necks expanded to more than twice the normal girth, their lyre-shaped tail feathers shiny and shaking, their bright white, powder-puff bum feathers held sharply past the vertical against the tail, their wings slightly lowered to expose two bright white shoulder spots against a black background. The roo-kooing is a repetitious musical phrasing that spells out an eight-note song: 'Hoo-doodle-woo-doodle-WOO-hoo'. I found an old bird watcher, Kenneth Richmond, rendering this as 'a stoop of sherry for Charrlie, a stoop of sherry for Charrlie', but I'm not sure that helps much. Edmund Selous's mnemonic is worse: 'Give him his coppers, he's going to take the electric'. Eh? Millais's comparison was with a goods train travelling over points. Maybe my rendition is not much use either. You have to be there, I guess.

The whole lek is conspicuous in the extreme. The sound of the sneezes carries for up to a mile on a still morning such as this and even the roo-kooing can be heard from a good way off, especially when six cocks are doing it at the same time. The white bums of the males, shaped like chrysanthemums, stand out even in poor light. They were how I found this lek, one early March morning a few years ago when I heard a faint whisper of Black Grouse sneezes on the wind and scanned a distant hillside till I saw some white spots moving about. At the time I was watching another, smaller lek that lies about half a mile to the south on the next ridge and attracts around nine cocks most years.

The white bums of males, shaped like chrysanthemums.

The bigger the lek, the louder and more visible it is, which probably explains why large leks tend to attract more females, perhaps disproportionately more, than small leks and may give a clue as to how lekking evolved in the first place. With females preferring to visit large leks, most males will have been fathered by a male at the centre of such a lek, so may have inherited an instinct to join a large one – despite the formidable odds he faces against winning the sexual lottery on that lek.

It was Rice University's David Queller who set out most clearly back in 1987 the argument that a female preference for aggregated males could drive the evolution of leks, even if it meant that most males got no advantage from joining them. Just as female choice can force a Peacock to wear a hefty train, so it can force a Black Grouse to join a lek: 'All it requires is that the genes causing males to display in larger groups also have a tendency to cause females to prefer to mate in larger groups.' Queller's mathematical models showed that this effect could evolve as easily as the exaggeration of individual plumage.

Yet big leks like these often attract Goshawks or Buzzards and once these raptors have learned where to come, they can devastate a lek. So could hunters in the past, and even today in parts of Scandinavia and Russia, shooting males at the lek in springtime is an occasional habit – and hazard for the males. Selous's lek in Sweden was disturbed by hunters several times, to his rage. Both sexes are taking a big risk in doing their mating this way. The Finnish scientist Ilkka Koivisto told of males cautiously returning to their territories even as a Goshawk was eating a female Black Grouse on a lek. Likewise, Carl Soulsbury, a scientist who studies the birds in Finland today, watched a Goshawk sit for a long time right in the middle of a lek. After a while most of the males nervously moved back on to the lek, desperate not to abandon their hard-won territories, while the astonished hawk watched his breakfast approach. The birds had to flee again when the hawk moved.

After Selous, other careful and objective accounts of Black Grouse leks began to appear. David Lack, a professional academic zoologist, watched leks in 1937 and 1938 in Scotland – just before heading to the Galápagos Islands to study Darwin's finches – and wrote detailed accounts of the birds' behaviour. He got to a lek at the unearthly hour of 3.40 a.m. on 13 April 1938 and saw the birds arrive at 4.25 a.m. He may have been the first to notice that each individual male has his own fixed territory that he occupies every day. (David Lack's DPhil student Chris Perrins was my DPhil supervisor so I am in a sense Lack's intel-lectual grandson, though I never knew him.)

A young Swiss-born medical scientist named Otto Hohn, who was an obsessive bird watcher, watched a lek in Staffordshire in England from 1941 to 1947 (when he emigrated to Canada and a career in avian endocrinology). The lek had been used for at least forty-seven years, according to local knowledge. It is rather poignant to read his remark, that 'at Hay Head leks, I found the birds already in display as early as 4.30 a.m., GMT, on April 20th, 1945' – that is ten days before Adolf Hitler shot himself. Wartime prevented lots of things, but not bird watching, it seems. Hohn became sufficiently adept at imitating the roo-kooing calls that he could induce the birds to join in. Unlike Lack, who thought the target of the roo-kooing was other males and the call aggressive, Hohn thought it was intended to attract females.

At Loch Ard in Scotland in the 1950s Kenneth Richmond crouched in his hiding place and found that he could chat, change a roll of film and light his pipe without disturbing the obsessed birds. He grew quite lyrical, writing of 'the tail feathers ashimmer with ecstasy' and young males being 'quickly worsted and sent about their business'. But he had strong and strange views about the function of a lek. Seeing no females show up, he concluded that the purpose of the ritual had been misunderstood, that males were displaying neither towards each other nor for the benefit of females, but to gratify and stimulate themselves. The primary function of a lek was social rather than sexual. To call it a dance was not entirely fanciful, he suggested. Quite how it would have evolved for this purpose – what survival or reproductive benefit it supplied – he did not explain. But, as I shall argue later, this was a time when Darwinism had got itself in a muddle and the theory of sexual selection had fallen out of fashion. Most likely, Richmond was visiting before or after the relatively short window – about two or three weeks – in which females mostly visit the lek.

By the 1960s lengthy scientific papers replete with actual data had begun to appear on the continent to replace the chatty and purple-prosed British natural-history narratives. Between 1960 and 1963, Ilkka Koivisto closely watched several leks in southern Finland, after capturing the birds and marking them. He made a clear distinction between three classes of male: the central, 'first class' males with small territories who attended every day and displayed almost incessantly; the peripheral 'second class' males with larger territories who displayed

A central 'first class' male.

less intensively; and the non-territorial 'third class' males that showed up from time to time. This is an accurate distinction on my lek too. Koivisto also recounted a rather tragic tale of a first-class male that was trapped, tagged and released in 1961, returned to the same lek in 1962, was trapped again to renew his damaged tag and released again. This time he took eight days to reappear on the lek and found his territory taken. He now had to accept that he was a second-class male on the periphery; he soon disappeared and was later seen displaying on a temporary lek with two other males. Koivisto also concluded that whereas males were somewhat weather-dependent in their display behaviour, females were ready to mate at the same time every year. As to the purpose of lekking, Koivisto thought it was all about predators:

by displaying in groups, male Black Grouse had more chance of not being ambushed by Goshawks. More eyes on the sky.

In the Netherlands between 1961 and 1964 J.P. Kruijt and Jerry Hogan watched two leks on meadows near to heather moorland in an area where the species is now almost extinct. For the first time the lek was subject to systematic recording, with the space divided up with numbered markers and minute-by-minute recording of each bird's behaviour. The birds' schedule was clear: males arrive about an hour before sunrise; females, if they come, turn up around sunrise itself. Copulation, if it happens, takes place in the first twenty minutes after sunrise. These two studies turned observation into quantitative data, at least for some of the activity on a Black Grouse lek.

'Full and round like strawberries'

Sitting in my hide today (with no desire to light a pipe), and watching the blackcocks dance and sing, I nonetheless try to forget my book learning and just plain observe what I see, to be as objective and unbiased as I can be. To be like Selous. The males are drawn to this place, of that there is no doubt. This becomes clear when all the birds suddenly fly off, frightened by some threat, possibly imaginary, or maybe real: a Buzzard or Goshawk may have flown past, unseen by me through the small side-windows in the canvas hide. When this happens the males fly as a flock, all heading in the same direction and land together about three hundred yards to the west, seeming to emphasise the sense that, despite all the fighting, the lek is a team effort by a group of semi-fraternal chums. I can see them standing tall and clearly watching the lek site, as if trying to pluck up courage to return. One then takes off and flies back, landing near the lek and running swiftly towards his territory. Immediately there is a rush to follow suit, and as the birds land short and run across the arena it's clear each one knows where he is heading. He seems in a hurry to get there and to resume displaying. Only once he reaches his miniature kingdom do his eye combs swell, does his tail fan out and his neck inflate. Inside the invisible walls of his little fortress, his cowardice is transmuted into courage.

The arbitrary formality of the lek is remarkable: each bird knows his place and probably has done since October. He seemingly experiences no feelings of aggression unless he is in it. Quite what happens when a bird dies and a vacancy arises, I'd love to witness, but males on average move steadily towards the centre of the lek as they grow to three or four years of age – their prime. So it must be a simple case of grabbing vacant real estate. Likewise cocks that fail to turn up each day must sometimes arrive to find their spot taken. I have yet to witness this, as Koivisto did.

The incessant displaying continues in the half-light. As Kenneth Richmond observed at Loch Ard, there does not need to be a female present – indeed there won't be one for another half an hour or more, if then. It's still early in the season and I may not see hens today at all. The males have occupied this arena every day for weeks, months even, without a single visit from their belles yet their faith is undimmed that Lady Godot will come today. In a typical season, females will visit the lek only on about three days each, mating on the third day. Almost all of the local females will come within about ten days, so this is a play with endless rehearsals and only a handful of full performances.

Undaunted by weeks of no-show from the targets of their affection, the cocks are spending almost all their time roo-kooing and sneezing. It must be exhausting. They take breaks mainly to pick fights with neighbours, only occasionally to pause and rest. Thus it was not unreasonable for the old naturalists to wonder whether the point of the display is to attract females. It might be to signal to rival males that you are on good form this morning and cannot be ousted from your territory without a fight; or maybe to simply enjoy the sound of your own voice and do what your hormones tell you to. That would be more an answer to the how than the why question, though.

The peripheral males, whose territories are larger since they have no neighbours on the outside, seem to spend less time displaying and more time peering hopefully over the grass towards the centre of the action. It's as if they know they don't have a chance this season but are here to learn and practise for next year. Sometimes they get ideas above their station: they take wing and land in the middle of the lek, only to be harassed and chased back to their second-class spots. These peripheral males are probably mostly one or two years of age. In their first breeding season, yearling males sometimes have patches of immature brown feathers on the back, head and neck.

They also grow smaller red eye combs. These eye combs are extraordinary things. An adult male on the lek is adorned with two lumps of swollen red flesh on his head – 'full and round like strawberries', as a German naturalist put it in the nineteenth century – that almost meet in the middle of the crown. Each comb is involuted into tightly packed tentacles about half a centimetre long. In order to count these tentacles, I took a close-up photograph one morning of a bird's head from about three metres away using an 800mm lens. It's so close you can see a reflection of my hide in the bird's eye. There are roughly 15 rows of tumescent tentacles on the comb, and at least 20 tentacles in each row, so more than 300 on each one. That's 600 stiff, tiny blood-filled and blood-red fingers of bare skin exposed to the breeze, a significant source of heat and moisture loss, presumably. From a distance they do indeed look like a pair of strawberries, but up close the resemblance to a pair of sea anemones is striking.

A study in Finland found that eye combs, which are shrunken and folded away altogether for the rest of the year, and indeed partly so for the rest of the day during the lekking season, start to expand and grow their tentacles under the influence of testosterone more than two months before the start of the breeding season. The combs also inflate during the display itself in a matter of minutes, pumped up with blood, then deflate fairly fast when a male stops displaying. Males two years old and older have larger combs than yearling males before and during the breeding season. In the Finnish study the heaviest yearlings had larger combs and the heaviest adults and those that had the most mating success had the largest combs of all.

A follow-up study by the Finland team, led by Carl Soulsbury, failed to find much of a correlation between the size of the eye combs and their brightness either in the red part of the spectrum or in the ultraviolet (the combs are bright in ultraviolet wavelengths too, a part of the spectrum we human beings cannot see but most birds can). So the bigger combs are not necessarily brighter. One year, on a different lek I watched, the alpha male that did most of the matings seemed to have a noticeably orange pair of combs; it did his strike rate no harm.

So if females or rival males want a clue to the vigour and strength of a particular cock, it seems they can get some idea from the size but not the brightness of the combs. Is that why they grow the combs – as an honest signal of their health and strength, a 'fitness indicator'? The argument goes that bright red skin features, which occur on quite a few male birds' heads, require the manufacture of carotenoid pigments, which are expensive and difficult to make. So it's hard to fake big eye combs. That's a key theory, and one I will revisit with scepticism in a later chapter.

But is the same true of the other ornaments? Do females or rival males read key information from the length and curvature of the tail, the whiteness of the bum, the brilliance of the white wing spots, the depth of blue sheen on the neck and back, the pitch and volume of the roo-kooing, the timbre and frequency of the sneezes? The idea that a displaying male is like a profile on Tinder, with each ornament or action giving a different detail about his prowess and fitness, begins to seem far-fetched. Surely if a male wants to advertise how fit he is, he could just grow as large as possible and thus intimidate every rival.

Why be such a fantastic and fancy fashion plate? Oh, and here's a funny thing: I find it hard to tell the males apart. Unless they have an injury or deformity, they all look alike and it is only from their position on the lek that I can name them. That's not true of the females. When they arrive some will quite clearly have more or different markings on their brown, neatly barred feathers. One hen is distinctly paler than the others. Another has fewer bars on her breast.

Just now, however, my reverie is interrupted by a practical task. It is getting light enough to see the whole lek and I'm going to try to count the males. Working from left to right, starting with a male who is almost behind the hide, I try to tick off each bird. There's one at the far side of the lek, who is not displaying, just watching, and I almost miss him. There are two threatening each other behind a clump of rushes and again I almost pass them by, but their sparring breaks into the open. The first count gets to twelve birds, but when I do it again, I get thirteen. I'm not sure which one I missed. Yesterday there were fourteen present so I have probably missed one, though peripheral males do take days off. Nor do I know whether I should count that watching bird at the far side. He might not belong to this lek; he might be a young bird scoping out the local leks, unsure which one to join: he does not seem to be displaying much. As the morning progresses, my counts will increase, swollen by the visits of these apprentice males.

2.

Darwin's Unpopular Theory

'Hope' is the thing with feathers –
That perches in the soul –
And sings the tune without the words –
And never stops – at all …

EMILY DICKINSON

5 a.m., April, the Pennine hills

I am certain that Charles Darwin never watched a Black Grouse lek. Yet he knew the species well. As a young man in the late 1820s he often went to the estate of his uncle (and future father-in-law) Josiah Wedgwood at Maer Hall in Staffordshire in August and September to shoot a few Black Grouse and Partridges for the pot, reporting his bags in letters to friends. 'Here are very few birds, I killed however, a brace of Black Game,' he wrote in September 1828 to his fellow beetle collector friend John Herbert. 'I returned from Black Game shooting at Maer on Monday,' he wrote in August 1829 to his cousin William Darwin Fox. Fifty years later, in 1879, he wrote to thank his friend George Romanes for sending him some Black Grouse to eat: 'I have not tasted Black Grouse for *nearly half-a-century*, when I killed some on my Father-in-laws land in Staffordshire.'

Though extinct there now, the birds would have been tolerably common throughout Staffordshire in Darwin's day. As late as 1860, the record British bag of 252 Black Grouse was shot in a single day on Cannock Chase in Staffordshire. But August was the only time of year when you could be sure never to see males at the lek even if you were an early riser: they would have been off in the woods and bogs and corn fields, moulting and feeding up for the winter. Darwin certainly knew about Black Grouse leks because in 1868 in a letter to Fox, now a vicar, he wrote that he had never heard of congregations of breeding birds 'except with Black cocks & some foreign birds'. In his book *The Descent of Man* in 1871 he quotes a detailed description of the Black Grouse lek, which he calls a 'balz' or 'spel', written by the German natural historian Christian Brehm.

The reason I am so confident Darwin never watched a lek is that he wrote in that same book that whereas these aggregations of male Black Grouse are common in Scandinavia and Germany, he had 'never met with any account of regular assemblages of black game in Scotland' – and by implication England either. A letter from one David Wedderburn sent on 6 April 1871, shortly after the book's publication, put him right. In Selkirkshire, Wedderburn had seen 'large numbers of blackcocks, in May, holding a regular "lek" on the grassy "haughs" by the river side'. Showing an infuriating lack of curiosity, Darwin simply altered the relevant sentence in later editions to read 'I have heard of only one instance, from Mr Wedderburn, of regular assemblage of black game in Scotland …' As other accounts attest, Black Grouse were lekking in most of the counties of England at this time, let alone Scotland, yet the country's most eminent biologist remained unaware of it, living as he did in one of the few counties where the species had never bred in living memory: Kent. This was not the only mistake Darwin made about the species. From Brehm he took the (wrong) lesson that 'the same black-cock, in order to prove his strength over several antagonists, will visit in the course of one morning several balz places'.

Standing tall for the lager-can sneeze.

More crucially he somehow imbibed the false impression that females do not attend leks, a fatal mistake. I wish Darwin had watched a lek because it would have strongly bolstered his extraordinary notion, which persuaded few of his contemporaries, that females may be the ultimate authors of males' extravagant plumage and behaviour in birds and many other animals. 'How interested Darwin would have been in these facts,' wrote Edmund Selous in 1910 after watching a Black Grouse lek closely, 'which, to my never-ending regret, were observed all too late for me to submit to him.' Drawing on Llewellyn Lloyd's account of Mr Hultman looking out of his bedroom window at females drawn to a caged cock, Darwin at least observed that 'the spel of the black-cock certainly serves as a call to the female' and that 'we are led to suppose that the females which are present are thus charmed'. But he went no further.

Female choice was a bold idea used to plug a hole in his theory of evolution by natural selection. Yet it had been born at least as early as its natural cousin. The historian of science Evelleen Richards has in recent years uncovered just how much, even in the 1830s, the young Darwin came to be almost obsessed with the idea of sexual as well as natural selection as a force driving evolution – and saw it as an entirely distinct idea. She suspects that the influence of his long dead grandfather Erasmus Darwin was acute. Erasmus was not just a (vastly overweight) physician, inventor, naturalist and poet of prodigious renown, but a free-thinking, free-loving liberal who openly fathered two illegitimate children and raised them alongside his nine surviving legitimate ones. His evolutionary poems narrowly escaped the charge of pornography by being ostensibly about the love affairs of plants and classical deities. 'Erasmus Darwin portrayed a universe governed by the sexual principle, by love and desire,' in Richards' words. 'This animal attraction is love,' Erasmus wrote, 'which constitutes the purest source of human felicity, the cordial drop in the otherwise vapid cup of life.'

In 1794 in his book *Zoonomia, or the Laws of Organic Life*, Erasmus Darwin wrote: 'The final cause of this contest amongst the males seems to be, that the strongest and most active animal should propagate the species, which should thence become improved.' It's a prescient harbinger of his grandson's theory. By April 1838, heavily influenced by his grandfather (a debt Charles played down for much of his life, living as he did in more puritanical times), the grandson was wondering whether 'species may not be made by a little more vigour being given to the chance offspring who have a slight peculiarity of structure. Hence seals take victorious seals, hence deer victorious deer.' In his 1844 essay, which formed the nucleus of *On the Origin of Species*, he wrote that 'the most vigorous and healthy males ... must generally gain a victory'.

If combat could winnow out the best, could beauty also? On 13 September 1838, two weeks before his momentous reading of (Thomas) Robert Malthus's essay on population, Charles again echoed Erasmus in his notebooks: 'The passion of the doe to the victorious stag, who rubs the skin off horns to fight, is analogous to the love of women (as ... seen in savages) to brave men.' The hint was born of female attraction working alongside male fighting to sort the virile masculine wheat from the feeble male chaff. During his adventures in South America as a scientist aboard HMS *Beagle*, Darwin had become intrigued by how it was that different human races, from Fuegians to Tahitians to Australians, sexually preferred their own kind. The diversity of human preference, such that women who seemed to him less alluring were clearly attractive to men of their own race, sowed a seed of an idea. If sexual preferences were different in different races, then people might look different as a result: that would be sexual selection. In South America he was attracted by Spanish ladies ('nice round mermaids') but still could not imagine preferring them to the perfect English angels he dreamed about, or so he wrote to his cousin.

Thus it was people, not birds, that first sparked a thought about selective sex as a force of evolutionary change. But birds were also on Darwin's mind. A few days after his rumination on human attraction, on 17 September 1838, with punctuation and syntax awry, he was wondering in his notebook about Peacocks displaying rather than just fighting: '(do the females then fight for male) & are merely most attracted). — singing best sign of most vigorous males. — (NB. most strange cocks & hens, being either alike or very different in recently altered genera. Guinea Fowl & Peacocks.!! other birds display beauty of plumage.'

'We must suppose Pea-hens admire Peacock's tail, as much as we do'

On 27 October 1838, two weeks before his wedding to his cousin
Emma Wedgwood, Darwin returned in his notebook to the nature of
beauty in racial terms:

> Consult the VII discourse by Sir J. Reynolds. — Is our idea of
> beauty, that which we have been most generally accustomed to:
> — analogous case to my idea of conscience. — deduction from
> this would be that a mountaineer takes [*sic*] born out of country
> yet would love mountains, & a negro, similarly treated would
> think negress beautiful, — (male glow worm doubtless admires
> female, showing, no connection with male figure) — As forms
> change, so must idea of beauty.

Darwin had suddenly immersed himself in the theory of aesthetics,
reading not only Sir Joshua Reynolds' *Discourses on Art*, but also
Edmund Burke's *Philosophical Enquiry into the Origin of our Ideas of the
Sublime and Beautiful*. From Reynolds he took the message that there
was such a thing as universal, timeless beauty, that it took effort and
study to appreciate it, that it was different from capricious fashion –
but also, and contradicting the first point about universality, that
'Ethiopian' races would indeed have different ideals of beauty. From
Burke he took the idea that beauty was essentially an erotic thing,
related at root to sexual attraction. But unlike Reynolds and Burke, he
began to wonder if beauty could be appreciated by other creatures than
educated Georgian gentlemen: by birds in particular, even females
ones. For birds, as for people, beauty seemed to be a mainly visual
thing. In March 1839 he was writing in his notebook: 'The tastes of
man, same as in Allied Kingdoms — *food, smell*. (ourang-outang),

music, colours we must suppose Pea-hens admire Peacock's tail, as much as we do.' There is the Peacock again, on its way to becoming the emblem, mascot, exemplar and epitome of sexual selection theory, as it would be throughout the next century.

In 1844 Darwin set out his idea of sexual selection in an essay on evolution. This essay would not see the light of day till 1858 when it formed part of the paper hastily presented alongside one from Alfred Russel Wallace to the Linnean Society announcing their independent discoveries of the theory of evolution by natural selection. Thus sexual selection was there as part of the theory of evolution from the very start. This is what Darwin wrote:

> Besides this natural means of selection, by which those
> individuals are preserved, whether in their egg, or larval, or
> mature state, which are best adapted to the place they fill in
> nature, there is a second agency at work in most unisexual [*sic*]
> animals, tending to produce the same effect, namely, the
> struggle of the males for the females. These struggles are
> generally decided by the law of battle, but in the case of birds,
> apparently, by the charms of their song, by their beauty or
> their power of courtship, as in the dancing rock-thrush of
> Guiana.

It was bantams that provided the next step in the argument. In 1842, Darwin had been in touch with an elderly Whig politician called Sir John Sebright MP, who bred chickens and had written a well-regarded book on animal instincts. This fellow, said Darwin, 'was an acute observer ... who bred all sorts of animals during his whole life, & who boasted that he could produce any feather in [three] years & any form in [six] years'. In around 1800 Sebright had generated a new and attractive breed of bantam, today called the Sebright, whose feathers are of dark gold or silver, each edged with black. The silver version looks like a drawing of a bird that is waiting to be coloured in. Curiously the male Sebright resembles a female, a fact that especially intrigued Darwin. In the first edition of the *Origin of Species* in 1859, he would give the idea of sexual selection just three paragraphs and summarise his argument thus, apparently with Sir John Sebright in mind:

> If a man can in short time give elegant carriage and beauty to his bantams, according to his standard of beauty, I can see no good reason to doubt that female birds, by selecting during thousands of generations, the most melodious or beautiful males, according to their standards of beauty, might produce a marked effect.

He was also thinking about pigeons. Darwin became obsessed with them in the 1850s, bothering pigeon breeders with importunate questions, joining two pigeon-fancying clubs and in 1855 starting his own dovecote at Down House to breed new crosses. He wanted to understand how one species of bird, the wild rock dove, had given rise to so many different varieties of 'fancy' pigeons: racers, tumblers, fantails, pouters and more. That these had come about as a result of selective breeding was key to his whole argument for natural selection. Pigeons so dominated the first chapter of his *Origin of Species* that his editor thought it was going to be a book about them. And perhaps females could breed fancy males by selecting the ones they fancied, in just the same way that pigeon breeders bred fancy breeds by selecting the ones they fancied as breeding stock. Indeed, his friend and mentor in the pigeon-fancying world, John Eaton, once wrote almost lustfully of his favourite breed: 'To my fancy, I am not aware there is anything ... so truly beautiful and elegant ... as the shape and carriage of the Almond Tumbler approaching perfection, in this property, (save lovely women [phew!]).' If a pigeon fancier can breed a tumbler pigeon, can a Peahen breed a Peacock?

This was a precarious argument to rely upon because if bantams and pigeons were being bred on purpose by people, did this not then imply that species were being bred on purpose by the creator? It was also a fairly flimsy scaffold on which to hang an explanation of something as universal and mysterious as beauty and it left Darwin feeling literally queasy. In April 1860 he famously wrote to the Harvard botanist Asa Gray that although he now felt he was winning the argument for natural selection and could explain the evolution of a complex organ like the eye to the satisfaction of most critics, 'the sight of a feather in a Peacock's tail, whenever I gaze at it, makes me sick'. By this he meant that ornaments on birds were a problem for his theory that loomed large in his mind, as the complexity of the eye had once done. If natural selection was a ruthless selective force, culling the unfit, then how on earth had it selected the gaudy, coloured train of the Peacock? This appeared to be a wasteful adornment, of no use in the survival of the bird – more likely a burden. And if the world was full of useless ornaments, maybe natural selection was not the convincing answer he thought. Darwin's bold new theory of natural selection appeared vulnerable to the charge that it was far too utilitarian to explain something as frivolous and futile as beauty.

The creationists by contrast had a ready explanation at hand for beauty: it was put there to delight mankind, an argument from natural theology that was all but unanswerable because it was circular. Sure enough, Darwin's critics soon fell upon exactly his inability to explain why beauty should matter. Throughout the 1860s his correspondence shows him wrestling with the issue of beauty in nature and reaching out to find people who might help him repel the theological argument. By the time of the fourth edition of the *Origin* in 1866, he felt obliged to add some paragraphs attacking the idea of beauty's purpose being to fascinate mankind. Sea shells were beautiful long before people were around to see them; flowers are beautiful for bees; berries for birds. So it made no sense to argue solipsistically that beauty is for us alone.

The hardest case, though, was bird plumage. 'I willingly admit,' Darwin conceded, 'that a great number of male animals, as all our most gorgeous birds, certainly some fishes, perhaps some mammals, and a host of magnificently coloured butterflies and some other insects, have been rendered beautiful for beauty's sake; but this has been effected not for the delight of man, but through sexual selection, that is from the more beautiful males having been continually preferred by their less ornamented females.'

Yes, but why would beautiful males be continually preferred by females? To this he had no answer yet.

'By a long selection of the more attractive males'

One of the most prominent critics of the *Origin of Species* was the eighth Duke of Argyll. While serving as Lord Privy Seal in Earl Russell's cabinet, this grand, sandy-whiskered, polymathic head of Clan Campbell found time to moonlight as a natural philosopher, indulging his passion for ornithology, anthropology and glaciology. He published a book in 1867, *The Reign of Law*, attacking Darwin's theory of godless natural selection in favour of a more theistic version of evolution. His most telling thrust came in a discussion of the different but still dazzling colours of different hummingbird species. What use in the battle for life is a frill ending in spangles of emerald rather than one ending in spangles of ruby, he exclaimed. 'A crest of topaz is no better in the struggle for existence than a crest of sapphire,' he added. Explain that by survival of the fittest! This was too much even for the novelist Charles Kingsley, a devout Christian, who wrote to Darwin expressing solidarity against the duke: 'What he says about hummingbirds is his weakest part … He has overlooked that beauty in males alone is a broad hint that females are meant to be charmed thereby.'

Yet Kingsley aside, to Darwin's contemporaries the answer that it all came back to female preference was also embarrassingly feeble and far-fetched. If females were picking fancy outfits for males, why? Most woundingly of all, his closest ally and co-discoverer of natural selection, Alfred Russel Wallace, was not persuaded. Why should the Peahen like a veil of eyes, and how did she acquire such a sophisticated aesthetic taste? Darwin was deeply disappointed when Wallace first told him in 1864 that he found the argument from sexual selection unconvincing.

If I am to understand beauty in birds, the development of Darwin's thinking during the late 1860s bears close examination, and the role of Wallace, as sceptic, is critical. They were mutual admirers and friends, Wallace having generously accepted Darwin's precedence as the author of natural selection despite the slightly odd way that the issue of priority for natural selection had been settled. Wallace's 1858 letter to Darwin from the Moluccas setting out the idea of natural selection had been read to a meeting of the Linnean Society in his absence, after a hastily written one from Darwin. Wallace was delighted rather than offended that Darwin had thus effectively achieved priority.

In 1868 Darwin, having finished his second book, on the variation of species under domestication, was preparing to write one defending his theory of natural selection against various criticisms, while extending the argument boldly into the history of humanity. He decided to devote nearly half of the new book, originally entitled 'On the Origin of Man' but eventually *The Descent of Man and Selection in Relation to Sex*, to the theory of sexual selection. The parallel between sexual selection and the deliberate selection of different breeds of chicken or pigeon was still uppermost in Darwin's mind. Just as people could endow 'the Sebright bantam with a new and elegant plumage', he wrote, so female birds, 'by a long selection of the more attractive males', have added to the beauty of male plumage. He thought mate preference was a big driver of evolution.

It is crucial to note that he saw sexual selection as quite distinct from natural selection. One worked through selective survival; the other through selective reproduction – sometimes at the expense of survival. It was a process of competition for mates, in 'battle or courtship, through his strength, pugnacity, ornaments, or music'. While sexual selection was generating beauty to help in mating, it might also produce handicaps to survival, by burdening males with ornaments that rendered them sexy but vulnerable to predators. So it was likely to be antagonistic to natural selection, at least some of the time. This was a point that future generations of biologists would struggle to under-stand in the following century: in the 1980s I recall giving a tutorial to a confident young student who adamantly refused to accept that repro-duction of the sexiest was different from survival of the fittest. And crucially – hence the inclusion in the new book – Darwin saw sexual selection as vital to explaining human racial differences. It appears that he could not see an explanation in natural selection for differences in skin, hair or body shape between human races, but influenced by his amazement that Fuegians and Australians found their own race attrac-tive, he wondered if differing sexual preference was the cause. At times it seems he also just wanted an excuse to write about sexual selection in animals and decided to tack it on to the new book. As he wrote to Wallace in March 1867, 'My sole reason for taking it up [the descent of man] is that I am pretty well convinced that sexual selection has played an important part in the formation of races, and sexual selection has always been a subject which has interested me much.' Sexual selec-tion was an entirely new idea, more so than natural selection, which had been hinted at by various naturalists before. It was his most origi-nal idea.

At first it seems Wallace did accept female choice as the cause of male beauty in birds at least. In his book *The Malay Archipelago* (published in 1869 but recording his experiences in the 1850s), Wallace would write: 'The successive stages of development of the colours and plumage of the Birds of Paradise are very interesting, from the striking manner in which they accord with the theory of their having been produced by the simple action of variation, and the cumulative power of selection by the females, of those male birds which were more than usually ornamental.' But he had begun to change his mind. On 17 March 1868, Darwin wrote to Wallace:

> I have been working hard in collecting facts on sexual selection every morning in London, and have done a good deal; but the subject grows more and more complex, and in many respects more difficult and doubtful. I have had grand success this morning in tracing gradational steps by which the Peacock tail has been developed: I quite feel as if I had seen a long line of its progenitors.

Wallace replied the next day:

> I do not see how the constant *minute* variations, which are sufficient for Natural Selection to work with, could be *sexually* selected. We seem to require a series of bold and abrupt variations. How can we imagine that an inch in the tail of a Peacock, or a quarter of an inch in that of the bird of paradise, would be noticed and preferred by the female?

To which Darwin responded tartly:

> In regard to sexual selection. A girl sees a handsome man, and without observing whether his nose or whiskers are the tenth of an inch longer or shorter than in some other man, admires his appearance and says she will marry him. So, I suppose, with the pea hen; and the tail has been increased in length merely by, on the whole, presenting a more gorgeous appearance.

This was surely an own goal. The idea that a woman falls in love with a pompous Victorian gentleman because of (rather than despite) his whiskers cannot have been very persuasive even at the time. He would have been better to stick with Peacocks. When I came across this letter it rang a bell and sure enough I soon tracked down a passage that Wallace wrote a quarter of a century later, in his book *Darwinism* (1889). He had waited a long time to deal with the whiskers:

> A young man, when courting, brushes or curls his hair, and has his moustache, beard, or whiskers in perfect order, and no doubt his sweetheart admires them; but this does not prove that she marries him on account of these ornaments, still less that hair, beard, whiskers, and moustache were developed by the continued preference of the female sex.

He went on: 'Female birds may be charmed or excited by the fine displays of plumage by the males; but there is no proof whatsoever that slight differences in that display have any effect in determining their choice of a partner.' Wallace allowed that bright colours might sometimes be useful to both sexes in ensuring that they mated with members of the right species, an argument that would be revived in the twentieth century by Julian Huxley. But this struggled to explain why in many species only males were brightly coloured, why some birds were dull-coloured in both sexes, and why males displayed their ornaments so vigorously. Just to observe a Peacock or a male Black Grouse is to see that he appears to be under no illusion that his inamorata is a member of the correct species, even though she lacks a gorgeous train or a fancy lyre tail to tell him so.

Wallace and Darwin both claimed support from the fact that beautiful colours and beautiful songs rarely go together. The calls made by Peacocks, lekking grouse and birds of paradise are disappointingly short, repetitive and ugly. Their purpose seems mainly to be to tell the females where to come to see the play. The endlessly inventive, varied and melodious songs of Nightingales, Song Thrushes, Skylarks and various warblers, by contrast, are performances in themselves, and they issue from the syrinxes of dull brown male birds that are all but indistinguishable from their females. This, Wallace thought, was evidence for his theory that display was just an excess of enthusiasm. But it fits Darwin's theory even better: the song of the Nightingale is an audio version of a Peacock's tail. No need to look flashy if you sound good.

'We may give the under surface to Mr Wallace, but we must yield the upper surface to Mr Darwin'

Bizarrely, as Helena Cronin spotted when writing her 1991 book *The Ant and the Peacock*, Darwin now also diverged steadily towards a less reasonable position on the colour of female birds in general. Wallace based his doubt about sexual selection partly on his view that the dull colours of females were actively chosen by natural selection to camouflage (a word that did not, however, enter the language until 1917) them on the nest. Darwin, instead of accepting this but insisting it could live alongside sexual selection of males, increasingly felt the need to reject it. In the fourth edition of the *Origin* in 1866, he had cited Wallace's argument and gave two examples of inconspicuous female birds: the Peahen, in which a long tail would be an encumbrance, and the female Capercaillie (or Capercailzie as it was then known), in which bold black plumage would give the nest away. In the sixth edition he removed the reference and in the *Descent* he described the notion that females had been specially modified as 'very doubtful'. By August 1868, towards the end of that year's correspondence with Wallace about sexual selection, he wrote that 'it has vexed me much to find that the further I get on, the more I differ from you about the females being dull-coloured for protection'.

Darwin's argument, at least in part, held that there was no evidence
to suggest the plumage of female birds was an arrested version of the
plumage of males: a Peahen's tail is not a Peacock's tail halted in its
tracks at the half-grown stage by predators. But this was a bit of a straw
man, a misunderstanding of Wallace's point. Cronin could not explain
to her satisfaction what else Darwin found objectionable about the
idea of females being selected for camouflage. It had much to do with
his muddled ideas about heredity and especially the problem of
sex-linked inheritance: female creatures giving birth to males without
passing on their own traits to them, or vice versa. Darwin subscribed
to a theory of inheritance called pangenesis, in which he imagined that
tiny replicas of each part of the body migrated to the embryo of the
new creature, and this made it hard for him to accept that natural
selection could operate on one sex, sexual selection on the other.

The notion of a latent or dormant influence, an unexpressed gene,
awakened in one sex by a hormone, so easily solves this problem, but
before genetics it was not available. Possibly also Darwin seems to have
felt that if he admitted selection to be involved in protecting females
from their enemies, it would surely do so for males as well, crowding
out sexual selection. Thus in a passage about fishes, he points out that
males are sometimes gaudy even when they look after the eggs – in
sticklebacks, for example.

During 1868 the disagreement between Darwin and Wallace over
sexual selection became a growing obsession with both men. In
September that year, Darwin wrote to Wallace to say that 'You will be
pleased to hear that I am undergoing severe distress about the protection & sexual selection: this morning I oscillated with joy towards you:
this evening I have swung back to [my] old position, out of which I
fear I shall never get.' That same month he invited Wallace to Down
House for a weekend, along with three pro-Darwin naturalists, in an
attempt to thrash out their differences. The two allies who came (Henry
Bates turned down the invitation) were John Jenner Weir, a customs

official and keen amateur biologist who was increasingly at Darwin's beck and call for observations and experiments, and Edward Blyth, the 'father of Indian ornithology' who had been curator of a natural history museum in Calcutta. Blyth was by now a somewhat tragic figure, impoverished, widowed and alcoholic, but he had been one of Darwin's most useful and prolific correspondents with a strong and early interest in evolution, albeit through a lens of natural theology.

No record of what was said that weekend at Down exists but the correspondence afterwards suggests the three of them ganged up on Wallace. They tried to persuade him that – given the dull colours of juvenile birds – it was more parsimonious to argue that the appearance of a few red feathers on the head of an adult male was due to sexual selection in males than that it was a preceding phenomenon subsequently deleted in females. Or something like that: the red feathers feature in a confusing exchange of letters after the meeting. Evelleen Richards reckons this September summit hardened rather than dissolved the differences between Wallace and Darwin. 'I fear we shall never understand each other,' wrote Darwin afterwards. Certainly, in their letters, the debate now petered out. As Richards puts it, 'By October, having depleted their respective stocks of arguments and exhausted their powers of persuasion, they agreed to disagree with mutual protests of sorrow and esteem.'

If only the two bearded scientists had been able to read a clever remark in a short letter in *Nature* from a butterfly collector, George Fraser, written three years later in 1871 in response to the *Descent of Man*. In the blue butterflies, Fraser pointed out, the under surfaces of the wings, exposed when at rest, are dull and difficult to spot: 'specially adapted for protective purposes', he reckoned. Whereas the upper surfaces in the males are bright blue and conspicuous, designed for charming females, he thought. Fraser adds: 'We may give the under surface to Mr Wallace, but we must yield the upper surface to Mr Darwin.'

'They are like skies where the stars are all moons'

Back in March 1868 Wallace had suggested to Darwin that he visit a young man named Thomas Wood, who had some insect pupae that might interest him. Wood was both a naturalist and a talented artist who had supplied Wallace with engravings of birds for his book on the Malay Archipelago. Darwin and Wood were soon in correspondence and in June 1870 Wood sent Darwin an article he had written that described in detail the courtship behaviour of birds, including the Black Grouse and the North American Heath Hen, a now extinct grouse then found on Long Island. Wood's eye for detail was as good as his sketching. Of Peacocks he wrote:

> This action of display is performed by the males, generally, though not always in the presence of females, and undoubtedly has for its object the winning of their favours. How intently is the attention of the Peacock fixed upon the Peahen when he stands before her with his glorious train of oscellated [*sic*] feathers fully expanded.

This was music to Darwin's ears. Yet even so, like his friend Wallace, Wood could not accept female choice as a selective force. Although he felt 'convinced of the truth of your theory of the origin of Species', he thought that bird beauty could not be explained by it, 'but seems to point to (& almost to prove) the existence of a great artistic power'. His trump card was a bird from the forests of Borneo, Malaya and Sumatra: the Great Argus Pheasant. This dazzling creature makes even a Peacock look pedestrian, because its immensely elongated, almost three-foot wing feathers are adorned with rows of circular, golden decorations, some 150 of them on each wing – so about twice as many ocelli on the whole bird as there are on a Peacock's train. These circles are so shaded as to look three-dimensional when held at a certain angle: 'like rows of convex and highly polished pebbles', as Wood put it. Or 'like the "ball and socket" ornament which is common in the decoration of human art', as the Duke of Argyll put it. Or as our old friend Edmund Selous put it many years later in a children's book: 'Shall I tell you what such

The ocelli on Great Argus wing feathers look spherical,
as if lit from above.

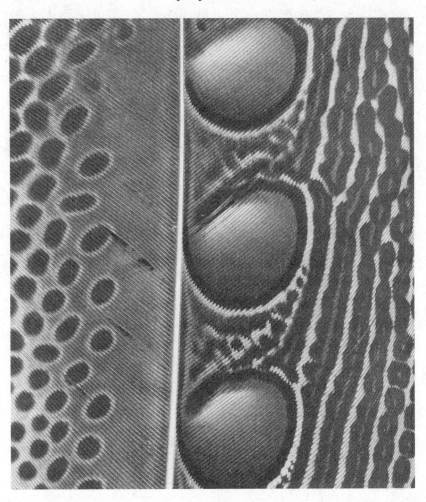

wings are like? They are like skies where the stars are all moons, that float softly among soft brown and amber clouds, tipping them all with soft silver.'

Indeed, as Wood explained to Darwin, the shading and highlights on the Great Argus ocelli were so accurate that each feather was subtly different. Some of the feathers were held at an angle during the bird's display and sure enough their highlight was off centre so as to be directly at the top of the orb when the feather was held that way. Only when displayed at the correct angle during courtship did all the balls or pebbles appear three-dimensional and lit from above. To Wood, such patterns 'present exactly the appearance of having been painted with consummate skill by an artist whose perfect knowledge of the laws of light had enabled him so to deceive the eye as to cause a flat surface to appear convex'. He thought it very difficult to believe that birds could appreciate the beauty they were gifted with, or that 'such a superabundance of the most beautiful ornamentation that can be imagined should grow out of the flesh of a poor bird, and be renewed annually, for the sole purpose of fascinating his lady love'. Surely a refined taste for the beautiful – and appreciation of how to shade something so that it looked three-dimensional – was a characteristic of only the highest civilisation?

It was not until 1872 that a live Great Argus Pheasant was acquired by London Zoo, from Singapore. But dead specimens reached London before that. After receiving Wood's letter in 1870, Darwin went to look at the British Museum's stuffed specimen. It had been mounted with the wings trailing downwards and he was greatly disappointed to see the circles as flat or concave: what was the engraver on about? Only when John Gould, the artist and collector, told him that the bird holds its wing feathers vertically upwards in display did the decorations look three-dimensional. In that position they were so shaded as to appear to be lit from above and shaded below. 'These feathers have been shewn to several artists, and all have expressed their admiration at the perfect shading,' wrote Darwin.

An anonymous reviewer of the *Descent of Man* would soon use this very point to mock Darwin's argument that the artist in this case was really generations of discerning female birds. Darwin had incautiously admitted that it might seem incredible that the full ball-and-socket effect of the Argus feather could result from tiny changes over many generations, none of which was intended to produce that effect – as incredible 'as that one of Raphael's Madonnas should have been formed by the selection of chance daubs of paint made by a long succession of young artists, none of whom intended at first to draw the human figure'. The reviewer seized on this, caricaturing Darwin's argument as: 'We must attribute to a hen Argus Pheasant the aesthetic powers of Raphael in order to account for the decorations of her mate.'

Yet the argument from incredulity has always been a futile one, for it applies just as well to the creator: the idea of a bird gaining the ability to appreciate beauty through natural selection may be hard to accept, but why is it any easier to accept the idea of a deity randomly implanting that ability in the bird? True to form, Darwin spent eight pages of his book showing how the ocelli on the Great Argus wing feathers had a continuous series of close cousins elsewhere on the feathers, tracing their origin back to simpler patterns. He devoted another seven pages to explaining the peculiar shape of the Peacock ocellus with its indented central disc. He triumphantly explained this after examining the ocelli on the tail coverts of related birds called Peacock Pheasants. In one species there are two ocelli on either side of the shaft of the feather; in another these two have partly merged; in the Peacock there is a single ocellus centred on the shaft but with an indentation to indicate a more complete merger in the past. Like a real eye, therefore, the ocellus was evolved by gradual stages, he argued, of which these other species or other feathers represent modern echoes. No supernatural explanation required, or conscious artistic expertise in the male or female mind as to how to represent or interpret three-dimensional illusions.

For later editions of his book Darwin commissioned Thomas Wood to draw a beautifully accurate engraving of the posture adopted by the male Great Argus Pheasant during its display, presumably based on the bird that arrived at the zoo in 1872 since the dance had never been described in the wild. When the male displays, the greatly elongated wings explode into a sort of cone aimed at the female, two long and floppy tail feathers showing over the top, and one real eye of the bird placed neatly at its apex. Darwin makes much of the fact that the spectacular wing feathers of the Great Argus remain hidden except during the display directed at a female, which is 'good evidence that the most refined beauty may serve as a sexual charm, and for no other purpose'.

Wallace disagreed. In Malaya in 1854 he had heard Great Argus Pheasants often in the forest but when he asked a local to shoot one for him, the man replied that 'although he had been for twenty years shooting birds in these forests he had never yet shot one, and had never even seen one except after it had been caught. The bird is so exceedingly shy and wary, and runs along the ground in the densest parts of the forest so quickly, that it is impossible to get near it.' What is more, it was actually well camouflaged: its 'sober colours and rich eye-like spots, which are so ornamental when seen in a museum, must harmonize well with the dead leaves among which it dwells'. It was this very species that 'first shook' Wallace's belief in sexual selection.

'A throbbing, shimmering hemisphere of three hundred vertically illuminated golden spheres'

In the wild the Great Argus dance happens deep in the hill forests of Borneo and the Malay Peninsula, on special dance sites cleared by the male tenant, a (tasty) bird trained by generations of hunters to be extremely wary. In the 1920s the explorer William Beebe eventually succeeded in seeing the male display briefly after a week of watching from perches in a tree or a hole in the ant-infested ground. Beebe wrote that he would 'prefer to accept' Darwin's theory, but could not. 'I should be willing to strain a point here and there to admit this pleasant psychologically aesthetic possibility; but I cannot.' Like Wallace, he found it impossible to believe that female choice had anything to do with the wings of the Argus, because hen pheasants can't do aesthetics: 'There is no question in my mind that the wonderful colouring, the elaborate ball-and-socket illusion of the ocelli, the rhythmical shivering of the feathers which makes these balls revolve – all are lost, as aesthetic phenomena, upon the nonchalant little hen.' As to the real reason the bird acquired its plumage: 'It is one of those cases where we should be brave enough to say, "I do not know."'

In the 1970s Geoffrey Davison spent many days watching Great Argus dance sites and saw males displaying many times and for the first time witnessed a full sequence of display when a female visited one site. In the 2010s, the dedicated bird photographer Tim Laman and his Bornean colleagues, Darmawan and Wahyu Susanto, captured the dance on film.

When at rest the male, not much bigger than a rooster, appears discreetly brown with pink legs and a blue, bare face, except for his massive wings, which are almost twice the length of his body, and beyond that his two immensely long tail feathers crinkled like the crest of a newt. He wanders about the dance site picking up leaves and twigs, occasionally issuing his loud 'Pow-wow!' call. When a female appears he rushes around her quite suddenly lifting his huge wings into a cone of shivering feathers that almost engulfs her as he rustles it forward repeatedly. His head turned to one side, he looks at her through the tiny gap at the apex of the cone. As ornithologist Richard Prum puts it in his book *The Evolution of Beauty*: 'Taken together, the elements of the male display add up to a sensory experience of mind-boggling complexity – a throbbing, shimmering hemisphere of three hundred vertically illuminated golden spheres that instantaneously appear

*Lifting his huge wings into a cone of shivering feathers
that almost engulfs her. Photo by Tim Laman.*

suspended in the air against a feathery background tapestry of speckles, dots, and swirls.'

(There is another Argus pheasant species, the Crested Argus, which lives at a higher altitude than the Great Argus in Malaya, Laos and Vietnam. Its wings are quite normal but it grows the largest feathers in the world in its tail: six feet long and more than six inches wide. It displays and breeds in captivity but has never been watched in the wild, where hunting and logging threaten to drive it extinct.)

In 2022 two scientists at the University of Exeter's Cornwall campus set out to test whether birds did indeed experience the illusion that Great Argus 'balls' are three-dimensional. They printed a series of twenty-six artificial images, half of which were shaded to look like convex spheres and half like concave cups. They then showed both to twenty-two chickens. When eleven of the chickens pecked at the convex images, they were rewarded with a mealworm; the other eleven were rewarded for choosing the concave images. The birds soon got the point and half became convex pickers, half became concave pickers. Next the chickens were presented with two photographs of one of the ocelli from a Great Argus Pheasant wing feather, isolated from background material and printed either right way up (which looks convex to us) or upside down (concave). Sure enough, most of the chickens trained to pick convex shapes chose the convex Argus picture, while most of the ones trained to pick the concave shapes chose the concave Argus picture. Probable conclusion: birds can see 3D illusions. However, Richard Dawkins in his autobiography *An Appetite for Wonder* tells how he did experiments in the 1960s that showed how even newborn chicks innately see 3D illusions if they appear to be lit from above.

Why would a male bird want to show optical illusions to females? Argus Pheasants eat fruit a lot, so perhaps shiny round things catch their attention. Like many prey animals they are easily frightened by the eyes of owls or cats, too. So a tendency to pay attention to spheres may be instinctive in such species. Sexual selection may have taken a pre-existing attention bias and exploited it. But – here's the problem with that theory. Eye-like and fruit-like objects play no part whatsoever in the display of the Black Grouse, the Ruff, or many birds of paradise. These 'sensory exploitation' ideas cannot easily explain the variety of bizarre colours and shapes that sexual selection appears to have exaggerated across the diversity of birds.

Great Argus Pheasants do not lek. The dance floor is occupied by a single, solitary male who aggressively expels any other male, while calling loudly to attract females to watch him dance. His nearest rival may be miles away. This is true of many forest-dwelling birds. The Capercaillie, a pine-forest specialist cousin of the Black Grouse, is said to have an 'exploded lek' in that males are in earshot but mostly out of sight of one another. Likewise, the Spruce Grouse, the Ruffed Grouse and the Dusky Grouse of North American forests all display at set sites, but not in close company with other males. This pattern seems also to be true of many colourful and ostentatious pheasants, from the gaudy Golden Pheasants of Chinese bamboo forests and the iridescent blue Monals of the high Himalayas to the extraordinary Bulwer's Wattled Pheasant of Borneo, whose dance turns the male into a sort of great disc of white feathers (from the tail), topped with a bizarre tube of pale blue flesh, made from tumescent wattles that swell to envelop his entire head. The tragopans of the Himalayas are a genus of forest birds in which the males are a gorgeous crimson dappled with black-bordered white spots. That's stunning enough but when displaying to a female, a male Temminck's Tragopan will hide behind a log as he slowly shakes loose a lappet of skin beneath his throat, expanding it into a smooth apron of electric blue with a zig-zag pattern of bright red. Two fleshy blue horns spring up behind his skull, pulled by muscles over a special knob of bone to raise them aloft. He then leaps out from behind his log, whirring his wings at the female, in what is clearly an attempt to stun her with the bright blue-and-red bib. But here's the problem. How does she compare him with another male that might be a mile away? In lekking birds it is easy to see that comparisons can be made; not so much in those that display in more solitary ways.

'Does the male parade his charms with so much pomp and rivalry for no purpose?'

In the *Descent of Man* Darwin devoted reams of print to accumulating myriad cases of colourful animals that had probably acquired their colours by sexual selection, not just in birds. The breadth and depth of his examples were truly astonishing. His technique by this stage in his writing career was to adduce instance after instance to support his theories, piling them up into a logical mountain till his opponents conceded defeat through mere attrition. From his numerous corre-spondents all around the world he gathered tales of beautiful males luring females with charm or prevailing over rival males with violence. There were newts with frilly crests, fish with bulbous snouts or elon-gated fins, frogs with ugly croaks, bellowing turtles, fighting chameleons, smelly snakes, battling beetles, beautiful butterflies and colourful bugs.

To single out just one group of animals from this vast menagerie, dragonflies and damselflies, Darwin noted, are nearly always sexually dimorphic: with brightly coloured males and dull females. Brushing aside the objection that insects are too primitive and simple to appre-ciate beauty, he included them among many other sexually dimorphic insects as likely cases of sexual selection. I reckon he is right, having often watched Banded Demoiselles by a stream near my home. The male of this damselfly, a bright, iridescent blue-green in the body, flut-ters and dances about over the edge of the water in a ritualised, almost slow-motion way, seeming deliberately to emphasise the large black bands on each of his wings to the bronze-green, plain-winged females. The thing about damselfly mating is that the female does indeed have a good deal of control over what happens. Once she allows the male to grasp her by the neck, it's up to her to bend her abdomen up and make contact with his penis, forming the famous 'wheel' posture. It's also up to her whether to mate with another male before laying her eggs, the last male to mate being generally the father of her young. So it's possi-ble, even likely, that Darwin was right and the bright colours and banded wings of the male Demoiselle are there to 'charm' the female.

Darwin went through scores of examples of mostly tropical butter-flies in which the males are far more colourful than the females and appear to employ those colours in courtship displays. 'I am led to believe that the females prefer or are more excited by the more brilliant males; for on any other supposition, the males would, as far as we can see, be ornamented to no purpose.' Likewise, he devoted several pages to describing the musical instruments that many crickets, grasshoppers and hemipteran bugs possess on their wings or legs. He pointed out that these stridulations seem to be used in alluring the females. 'All observers agree that the sounds serve either to call or excite the mute females.'

As for birds, to which Darwin devoted the majority of his argument, despite a huge weight of examples of males having brighter colours and displaying them to females, the direct evidence he could find for active female choice remained thin. He interrogated the most 'careful and experienced observers' among his correspondents for examples of female birds preferring one male to another. He received 'long letters' from two of them, 'almost an essay' from another, but the conclusion was disappointing: 'They do not believe that the females prefer certain males on account of the beauty of their plumage.' Indeed, one of these correspondents, a bird breeder named William Tegetmeier, who was usually accommodating to Darwin's requests, proved reluctant even to try the experiments Darwin suggested, such as dyeing a white male pigeon magenta and seeing if females liked him. Tegetmeier simply did not believe in female choice.

Darwin seized rather desperately upon a single somewhat ambiguous example, a story told by Sir Robert Heron, a Whig MP and baronet from Lincolnshire who bred Peafowl as part of an extensive menagerie. Heron recounted that his Peahens 'have frequently great preference to a particular Peacock', and one year their favourite was a pied cock who was confined in a cage, so the females 'were constantly assembled close to the trellis-walls of his prison, and would not suffer a japanned Peacock to touch them'. The next year, with the pied piper shut up in a stable out of sight, the Peahens happily accepted the japanned male – a variety with black shoulders. It was flimsy evidence of females choosing.

That Peacocks and birds of paradise display during courtship of females and not during fights with other males was one of Darwin's strongest cards. He asked: 'Does the male parade his charms with so much pomp and rivalry for no purpose?' If the most charming males are accepted by the females, then 'there is not much difficulty in understanding how male birds have gradually acquired their ornamental characters'. He resorted to a science-fiction analogy: 'If an inhabitant from another planet were to behold a number of young rustics at a fair courting a pretty girl, and quarrelling about her like birds at one of their places of assemblage, he would, by the eagerness of the wooers to please her and to display their finery, infer that she had the power of choice.'

'I remain firmly convinced of its truth'

When Alfred Russel Wallace came to review the *Descent of Man* in the periodical *The Academy*, he was at pains initially to praise his friend's examination of sexual selection. Many people, he pointed out, had observed how few paragraphs Darwin had devoted to this idea in the *Origin*, concluding that it was 'but a vague hypothesis almost unsupported by direct evidence'. Not so! The great evolutionist had now produced five hundred pages of exhausting detail to support his case and Wallace was impressed. It revealed a new world of animal life in which 'the structure, the weapons, the ornaments, and the colouring of animals, owes [*sic*] its very existence to the separation of the sexes'. He considered it 'one of the most striking creations of Mr Darwin's genius, and it is all his own'. Darwin was mightily relieved. 'The way he carries on controversy is perfectly beautiful,' he conceded to his daughter, though fearing that there was a sting in the tail.

Sure enough, Wallace's praise presaged a significant reservation. He could easily see how males could acquire weapons and pugnacity to win females with, but he balked at allowing those females the taste and the agency, let alone the motive, to select beauty and ornament. Surely, a capricious taste for colour among females 'would necessarily lead to a speckled or piebald and unstable result, not to the beautifully definite colours and markings we see'. Wallace continued to think that females were being ruthlessly selected by natural selection to be camouflaged, and somehow he thought this meant that males could not be being selected to be colourful by a different agency, mate choice. At least I think that is what he was arguing and it seems to be what Darwin thought he was arguing: 'Mr Wallace believes that the difference between the sexes seems to be due not so much to males having been modified, as to the females having in all or almost all cases acquired dull colours for the sake of protection,' he had written in the *Descent*.

Wallace's reservations provided ammunition to a much more belligerent reviewer. St George Mivart was a lawyer turned philosopher who had initially favoured Darwinism and had fallen badly out of love with it while converting to Catholicism. In a long and influential review he rubbished the *Descent*, appealing to Wallace's criticism to undermine Darwin's arguments and portray the author as (Darwin felt) arrogant and odious. Accusing Darwin of 'strangely exaggerating' sexual selection, he found 'no fragment of evidence' that 'female caprice' lay behind male adornments. He pointed out that female choice could not be true because Peacocks always make the first advances (half true, but females easily reject them if they wish). And like Wallace he considered female preferences far too unpredictable: 'Such is the instability of a vicious feminine caprice, that no constancy of coloration could be produced by its selective action.' As for human beings, it was nonsense to suppose that men of different races had different ideals of beauty. All cultivated Europeans, after all, agree on the 'Hellenic ideal'.

In the months and years that followed, few of Darwin's allies came to the defence of female choice. Thomas Henry Huxley, the biologist famous as 'Darwin's bulldog', was happy to go into battle against Mivart in his usual pugnacious way, but on other issues, not on sexual selection, a topic he avoided throughout his life. Darwin added material to the *Descent* in later editions, especially to insist that Mivart was wrong about female choice being capricious and inconsistent, but he could not provide much hard evidence. Mivart had clearly got under his skin: in 1873 he even contemplated suing Mivart over a vicious attack on Darwin's son George for writing an article in favour of easier divorce.

Nor was Herbert Spencer, famous champion of natural selection, prepared to defend Darwin's idea of sexual selection. In relation to the origin of music, Spencer was unimpressed by Darwin's argument that it began with male animals trying to charm females. 'Swayed by his doctrine of sexual selection [Darwin] has leaned towards the view that music has its origin in the expression of amatory feeling,' he later wrote. This was clearly nonsense because Spencer had heard a lark sing in January when it was well known they do not mate till March. So Darwin's view was 'untenable' and the true cause of bird song was 'overflow of energy', just as was the case with the 'whistling and humming of tunes by boys and men'.

How much of the hesitation of Darwin's friends to support him permeated in from the increasingly strict assumptions of late Victorian society about human mate choice it is hard to say. But it's not implausible that several of these respectable gentlemen felt a certain reticence about approving an idea based on the notion that females could be the ones deciding in favour of sexual liaisons – perhaps lest their wives found out that they thought this way. Darwin's publisher got cold feet about some of the passages in his book and would not allow the word 'sexual' in the title, although 'sex' was allowed. As if to prove the point, early feminists including the novelist George Eliot – a friend of Darwin – seem to have taken some comfort from Darwin. Her masterpiece *Middlemarch* was published the year after the *Descent* and its portrayal of a woman regretting her choice of husband is generally reckoned to be partly influenced by Darwinian thoughts. Yet to the vast majority of late nineteenth-century intellectuals, sexual selection was Darwin's slightly embarrassing idea that ought to be quietly forgotten.

3.

The Females Arrive

In the spring a fuller crimson comes upon the robin's breast,
In the spring the wanton lapwing gets himself another crest,
In the spring a livelier iris changes on the burnished dove,
In the spring the young man's fancy lightly turns to thoughts of
love.

ALFRED, LORD TENNYSON

5.30 a.m., April, the Pennine hills

It's light now and the lek will soon be bathed in the first golden rays of sunlight through a break in the clouds, a hint of relief to my frozen fingers and toes. The activity is now constant and frenetic. Each male is either fighting or displaying, or torn between the two. There is bustle and tussle, noise and poise, and it is fast and frantic. The light shines through the white feathers that stand proud above the rumps of the males, so that each bird seems almost to have switched on a chain of bright white lamps across his back. The contrast is acute with the blackness of the tail and body, somehow blacker now that everything else is lit up. The blueness on the neck seems defeated in sunlight, being at its best in the dull light of an overcast day or – as now – just before the sun rises. But on a bright day like this, with the dead grass beginning to glow blond in the swelling light, the black and white show is bold, conspicuous, even somehow artificial.

Suddenly a spasm of more than usually infectious excitement runs through the arena. The males begin a frenzy of flutter-jumping, the vibrato versions of their sneezes coming quickly one after another. 'Tshwewewewe'. They leap as they flap, sometimes twisting in the air and landing on the same spot. They have spotted an approaching female long before I did. She is flying in from the north directly towards the lek. She lands a little way short and stands watchful for a few minutes at the edge. In Scandinavia, where leks are usually surrounded by trees, the females will often perch atop a tree at this point, to get a better view. Here, I have had females fly up on to the top of my hide, their claws scratching against the canvas inches above my head. But usually they land outside the lek, stand alert for about five or ten minutes and then walk in towards the centre. Today, the male closest to the arriving female, a junior peripheral cock, approaches her, roo-kooing furiously. She stands, apparently indifferent, looking over the rest of the field. She preens a few feathers, as if to emphasise her nonchalance, then pecks at a grass seed head. Then she begins walking towards the middle of the lek, the young male running slightly to try to get ahead of her, his head held low, his throat swollen and pulsing, his tail spread. She walks past him.

She is approaching the next male's territory, the one I call Touching Combs because the strawberry-anemones on his head are so big they touch in the middle of the crown. He is displaying as close to the northern boundary of his court as he can get, facing directly towards her. His posture is different from the normal roo-kooing one. He is pressed extremely low to the ground, his legs invisible. He is holding dead still in this position, whereas routine roo-kooing often involves some moving about. But he is producing the same sound. This low-to-the-ground position was called the crouch by the ornithologist David Lack, who first commented on it, and you only see it when a hen is

The female stands watchful for a few minutes at the edge of the lek.

approaching a cock's territory. Elsewhere on the lek, other males are also adopting the crouch, facing towards the female, as if willing her to come in their direction. Some that are further away are sneezing and flutter-jumping, but they too seem to be looking to the lady.

She moves on into a third territory, Wonky Tail's, who breaks off from the crouch posture to strut rapidly in a wide circle around her while roo-kooing. He gets his name from the way his lyre-shaped tail feathers are always at an angle, leaning left, due, I suppose, to some slight spinal deformity. This makes him unmistakeable and he is back in the very same spot for the third year, just right of the lek's centre: close but no cigar as far as mating opportunities are concerned.

The circling is yet another ritual display, different from the crouch and the normal roo-kooing in that he now has his primary wing feathers spread so that some of them trail in the frosty grass, while his breast and belly feathers are lowered like aircraft undercarriage as if to emphasise his bulk. Seen from the front he now has not one or two but three bold, white spots on each flank, the lowest being part of the white wing stripe. He seems to be trying to manoeuvre round in a circle and eventually position himself sideways-on in front of her. But she turns aside. He tries again, circling her at a distance of a few feet. The effect is to show her every part of his plumage from the tumescent, red combs and the swollen, blue neck to the white spots above and below the shoulder, the white stripe on the primary wing feathers, the black curl of the lyre tail and the erect feathers of the white bum.

The male circles closer and closer, in smaller and smaller rings.

Someone who has never watched a lek before almost always comments at this stage on the female's apparently magnificent indifference, I find. It is almost insulting how little she appears to rate his valiant efforts. Where is the applause? But then since birds don't do facial expressions, let alone clapping, I am not sure how we would know whether she appreciates the show. Her heart might be going pitter-pat for all we know. Her very stillness suggests a certain fascination. She often raises her head high and moves it about as if trying to survey the whole lek. The impression that she is trying to work out which male is the central Adonis is hard to avoid, but maybe that's me reading too much into it. Besides, the whole point of mate-choice theory, as proposed by Darwin, is that she has evolved into a discerning connoisseur who is not easily impressed. If she were bowled over cheaply, or discombobulated quickly, she might not get the very best male. Generations of breeding have made her into this flinty-eyed, cool customer.

Black Spot has a small black spot on the left side of his white bum.

She moves on again and now she is in the territory of this year's top cock, Black Spot, identified by a small black spot on the left side of his white bum, and two small black spots on the right side. How do I know he is the top cock? Because yesterday he mated two females just beside that very tuft of rushes where she now stands. She stops and he circles closer and closer, in smaller and smaller rings, roo-kooing with trailing wings and lowered undercarriage. But she shies away from him when he gets too close, as if still unsure or even a little afraid, and for good reason. There is some roughness if not violence within the sexual act in quite a few bird species. And the whole experience is a bit novel for her. It may be only her first or second visit to the lek this year and if she is a yearling hen she will not have seen males behave like this before. Males come here every day for months on end; she visits on only three days on average, and she is about to take the plunge and mate, an act that will happen just once each year for most females. So she shies away from him: it's a big decision. His circling continues.

Then she moves out of his territory again as if keen to re-sample the wares on offer elsewhere in the lek, visiting the territory of Black Bar, a male with a thin black line across the top of one of his white bum feathers. But she soon returns to Black Spot's court. He raises his tail, his game and his hopes. Again he circles with trailing wings, getting closer and closer. After some minutes of this, she suddenly squats before him. A clear invitation. He moves close enough to place a tentative foot on her back, but when that happens she jumps away. This happens again and again. Conflicting impulses are still competing for dominance in her head, it seems. Just when I think she is about to let him mate, he suddenly breaks off to have a quick fight with Black Bar, who has trespassed over the border of his territory slightly. This provokes what looks to me like world-weary irritation in the object of his affection: why do boys prefer Mars to Venus, she seems to ask.

Incidentally, these distinguishing small black blemishes on the white feathers of the bum may indicate flaws in the fitness of males. Carl

Soulsbury and his colleagues found in Finland that black spots at the tips of feathers were more frequent in birds that mated less often and survived to the next season less often, indicating that they had suffered more stress at some point. Birds with black tip spots had often done more fighting in previous years, 'suggesting high investment in fighting leads to carryover effects on male condition'. Black Spot is not perfect, therefore, but it seems he is the best on offer.

Another female has landed on the west side of the lek and is progressing towards the centre in much the same way, attracting enthusiastic attention from each male as she passes. The sneezes and the fighting stop when she gets close to a male and roo-kooing becomes continuous. She enters Black Spot's court now. The first female takes offence at this and flares her tail, charging towards the newcomer, asserting her priority. The second female moves a short distance away but soon returns. There's only limited antagonism between females, it seems, but there definitely is some. I was surprised by this when I first witnessed it because it seemed to be at odds with the general point that this is a highly polygamous species in which a female shares her mate with many others. And it does not seem to occur in other lekking grouse species such as Capercaillie or Sage Grouse, where a crowd of eager females often surround the top male as he continues to display.

Black Spot circles again and again around the first female, his wings trailing in the grass, and then she squats once more and this time lowers her wing slightly, revealing bright white shoulder spots similar to the male's. In one swift movement he climbs on to her back, flapping his wings for stability, grasps the nape of her neck with his beak and bends his tail against her upturned vent. Their cloacas kiss (sexual selection is probably responsible for the loss of penises in modern birds, see chapter 8, but I digress). Sperm is transferred in an instant: mating takes about two seconds. It has to be fast because during the act he is – almost every time at this lek – pounced upon by up to three of his neighbours in a melee of chaos. On this occasion both Wonky Tail and

Black Bar plunge into the fray, trying to knock him off her back. It looks like they are either trying to prevent him copulating, or to insert their own ejaculation into the action. Escaping this fight with the loss of a few feathers and a little dignity, the female stands and preens around her vent a little, a habit that might be about making sure the sperm goes in, or might be more about checking she has not been infected with lice by her promiscuous partner. Perhaps it's just about repairing dignity. She then flies off, clucking gently, to feed in a meadow down the hill. The male returns immediately to roo-kooing.

That's it: two or three seconds of presumably pleasurable sensation for the male as the culmination of months of singing, dancing and fighting. As for the female that will be her only moment of sexual ecstasy, if that is what it feels like, all year. Many birds mate before

The pile-on as three males try to disrupt a fourth's mating.

laying each egg, but female Black Grouse usually acquire enough sperm in a single mating to fertilise all the six to ten eggs they will then lay over the next ten days or so. They usually visit the lek for two or three days before they mate, but rarely return after they have mated, unless they lose their first clutch of eggs to a badger or a fox.

I have tried to describe this one female's journey through the lek, but I'm not the only one who has seen this drawn-out indecision on the part of females. Here's a second-by-second account of another episode, taken from the meticulous Dutch study by J.P. Kruijt and Jerry Hogan in the 1960s: 'April 28, 1962 – 4.40 Female 1 lands between B and H; 4.41 to C, crouches; 4.45 to D; 4.46 to E, crouches; 4.50 to D; 4.52 to B; 4.54 flies to E; 4.55 to D, crouches; 4.56 to G, crouches; 4.59 to boundary between B and H; 5.00 to C; 5.01 to G; 5.02 to E; 5.03 to D, crouches; 5.11 to E; 5.12 to B; female 2 arrives on the lek, lands in C; female 1 flies immediately to C, crouches and mates; 15 seconds later female 2 mates with C. Both females fly away at 5.20.' It seems female 1 was very undecided between males C, D and E, but her mind was made up when female 2 arrived and went straight to C.

'I claim actually to have seen that which Darwin believed must take place'

Nowhere in the animal kingdom is the autonomy of female agency so clear. She, not he, decides when mating will happen by squatting and then holding her squat. She walks through the lek to the appointed bush untroubled and untouched, apparently at no risk of being forced into mating. Very occasionally I see an attempt by another male to rush at a female and force the issue, but it never seems to succeed. Only when she is ready, usually on the third day that she visits, will she decide to mate. This is no patriarchy – a fact which surely vindicates Darwin, who insisted that females were capable of choosing and that male display was all about influencing female choice, and not Wallace or many of Darwin's successors, who saw it in much more Victorian terms, as males intimidating rival males and fighting for the right to mate. So score one for female choice. But we are a long way off proving that such choice is responsible for the evolution of bright colours and fancy feathers.

The second female is still at the lek and soon she approaches the centre again to go through the same dance of indecision among the central males while Black Spot, Black Bar, Touching Combs and Wonky Tail each in turn circles round her. Eventually she too squats in front of Black Spot who scores his second mating of the morning. Then, to my surprise, a third female selects Wonky Tail. At last, in his third year on the lek, he gets to mate! Only something seems to go wrong. The wonkiness of the tail somehow prevents a proper cloacal kiss: she moves off and fails to preen her rear end as she would after a successful mating. Five minutes later she returns to Wonky Tail and solicits a second attempt, this time more successfully, despite the violent interference of Black Bar. I watch Wonky Tail closely for signs of triumph, relief, joy or anything to indicate that the end of his long wait to mate has affected his mood. But he simply resumes roo-kooing, picks a short face-off with Black Bar and lets out some hoarse lager-can sneezes. Black Spot gets most of the matings this year, but not all.

By what criteria did each of those females choose her mate? Was it the beauty of Black Spot's plumage or the vigour of his display? Or merely his position at the centre of the lek? Or was it possibly just that other females chose him?

As I watch the males strut their stuff, it strikes me that – despite my names for them – there is very little to choose between them to the naked eye, like men in tuxedos. Every male has bold white wing spots on each flank, a soft white bum erect at the rear (actually more than erect, tilted forward at forty-five degrees), an elegant black lyre-shaped tail to either side, a deep blue tint on the neck and back, two vermillion anemones on the crown of the head. If these features are like entries in a dating profile, delineating the quality of the males, they don't seem to be very informative: every male looks perfect, so why do all the hens mostly agree that just one of them is best?

My eye is drawn to a central male just to the right of Black Spot's territory, for whom I have not found a distinguishing mark and coined a name. Today, like yesterday, he has spent much of the morning half asleep, his head sunk into his shoulders and his red combs shrunken. He looks unwell. He has not done his share of the work attracting females to the lek. He woke up when the females arrived, flutter-jumped well and roo-kooed furiously, then picked a fight with Black Spot. So he is capable of holding his own, but his frequency of display over the whole morning has been far lower than his neighbour's. If the females detect his exhaustion they will surely pass him by. Is that why they stand and watch the whole lek from the edge?

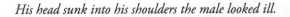

His head sunk into his shoulders the male looked ill.

That females choose, unmolested by males, was first spotted by Edmund Selous. Watching Black Grouse in Sweden in 1907, he found Darwin's faith in female choice vindicated: 'The general principles governing the courtship of the blackcock are much the same as those which obtain in the case of the Ruff – that is to say, the female is won by assiduity or superior attractions, and not by fighting, or force applied to herself.' And it's active choice on her part, not a choice being made for her. She shows 'no sign of that passive surrender which naturalists who are opposed to the doctrine of sexual selection talk about'. Instead, 'the power of invitation, permission, and veto, seems to lie with her'. This is exactly what I see with Black Grouse today. I wrote the following sentence in my notes long before I read Selous yet they could be his words. 'The hens move slowly through the lek, passing within territory after territory and being courted but not harassed. They seem wholly unhurried, perhaps even unimpressed – and most certainly capable of choice. The squat with lowered wings that precedes mating is an unambiguous sexual solicitation.' Some years later, when I took the primatologist Isabel Behncke to see the lek, she was amazed by this. 'You would never see this in mammals: a fertile female able to wander at will from male to male without fear of assault.'

On the Black Grouse lek he watched in England in 1908, Selous at last witnessed mating. He lamented the failure of Darwin's correspondents, who had told the great man 'only of the frantic part' of the male Black Grouse's behaviour so that Darwin had assumed this pugnacity was directed at impressing the female, missing the crucial detail that the fighting actually dies down when the females arrive. As I shall show later, this crucial detail observed by Selous – that fighting dies down when females turn up – has since been catalogued in other lekking species and is a vital clue to what leks are all about.

If only Darwin had known that she alone decides whether and when mating will occur – how excited he would have been! 'I claim actually to have seen that which Darwin believed must take place,' wrote

Selous. Deep within an obscure journal of natural history called the *Zoologist* that few people ever read, therefore, a bird-watching amateur was proving Darwin's point: 'How do the above facts affect the question of sexual selection? In my opinion, they speak, trumpet-tongued, in its favour,' he wrote. 'There is simply no place for the display of the male blackcock before the female, if it be not for the purpose of winning her, and my notes show he does win her.'

Selous was almost completely ignored. In 1913 he wrote a furious note in the Yorkshire publication of the North of England Natural History Society, the *Naturalist*, protesting at the continuing misinformation being written about Black Grouse. In the latest edition of *The British Bird Book*, he pointed out, his own observations had been missed in favour of older and wrong stuff about the males engaging in a 'species of tournament', where they fight for dominance, the 'vanquished' retreating and the 'victor' getting together a 'harem' of females before departing 'each to his own ground'. This string of tired clichés completely missed the role of male display followed by female choice. 'This silence,' thundered Selous, 'from whatever cause proceeding, is not in the interests of truth, and moreover does great wrong to Darwin, whose brilliant and most original theory of sexual selection my observations most strongly confirm.' Why, he complained, was Darwin's notion of female choice, so obviously true 'yet still so constantly denied'? It was a good question. (Selous would have been infuriated by the BBC's *Wild Isles* in 2023.)

'As utterly incredible as my views are to you'

We left Darwin and Wallace disagreeing with increasing rancour about bird beauty. By 1877 in an article in *Macmillan's* magazine, Wallace was making the laughably random case that colourful male plumage was 'due to the greater vigour and activity and the higher vitality of the male'. This was both a circular and a misogynist argument: that males have more vitality than females. And the brightest colours were attached to the most elongated and elaborated feathers or structures, including the Argus wing feathers, he said, not because these were used in courtship display but because that's just a law of growth: the more something grows, the more likely it is to be a bright primary colour. The alternative, of incremental elaboration of the Argus ocelli by tens of thousands of female birds all sharing a preference in one direction, was 'absolutely incredible'. Darwin replied in kind: 'That the tail of the Peacock and his elaborate display of it should be due merely to the vigour, activity, and vitality of the male is to me as utterly incredible as my views are to you.'

By the time he wrote his book *Darwinism* in 1889, Wallace had retreated from even his lukewarm acceptance of 1871 that display was intended to charm females. 'The extraordinary manner in which most birds display their plumage at the time of courtship, apparently with the full knowledge that it is beautiful, constitutes one of Mr Darwin's strongest arguments,' he conceded. But, he went on, it might be just an excess of vitality rather than seduction: 'Under the excitement of the sexual passion they perform strange antics or rapid flights, as much probably from an internal impulse to motions and assertion as with any desire to please the females.' As a young researcher into animal behaviour I was warned not to be anthropomorphic, but this is going far too far in the other direction. You would have to be blind not to see that the whole end and purpose of a Black Grouse's dance is to copulate.

Wallace was insisting that female choice was 'now no longer tenable' as a theory. Despite his book's title, *Darwinism*, he had come not to praise Darwin but to bury him – on this issue at least. Wallace played his trump card, or so he thought, and in doing so gave rise to what remains a highly popular theory to this day. Because growing long feathers, adopting gaudy colours or engaging in energetic displays was easier for healthy males, it followed that if these males were to be more successful in mating, the whole system would in effect be a form of natural selection, a method of eugenically selecting the best parents for the next generation. 'As all the evidence goes to show that, so far as female birds exercise any choice, it is of the "most vigorous, defiant, and mettlesome" males, this form of sexual selection will act in the same direction (as natural selection), and help to carry on the process of plume development to its culmination.' Sexual selection, if it operates at all, is therefore just a special case of natural selection, or at least is drowned out by survival of the fittest who happen to be brightest-coloured. All Darwin's appeal to capricious female agency can be dropped. Sexual selection had become needless, because 'natural selection which is an admitted vera causa will produce all the results'. Since Wallace's objection to sexual selection by female choice was in the service of natural selection, he was being more Catholic than the pope: 'Even in rejecting that phase of sexual selection depending on female choice, I insist on the greater efficacy of natural selection,' he crowed.

The vitality theory was still central to Wallace's argument. Bright colours, he asserted, are the natural end state of biological bodies because ... well, this is where things get a bit woolly. 'Colour may be looked upon as a necessary result of the highly complex chemical constitution of animal tissues and fluids,' wrote Wallace, vaguely praying in aid of the colour of blood and bile. If not checked by natural selection, he thought, every creature would steadily become more brightly hued and especially the males because males have this greater 'surplus of vital energy'. Because of the need for camouflage while sitting on the nest, natural selection is more active in 'repressing in the female those bright colours which are normally produced in both sexes by general laws'.

I'm staggered by this. Unlike Darwin, Wallace had watched leks – of birds of paradise. He knew as much of the beauty of male birds as anybody living. As a collector who sold stuffed tropical birds by the thousand for a living, he had made a special point of devoting five years to tracking down males, females and juveniles of each and every species of bird of paradise on the island of New Guinea. (They fetched good prices because of their fantastic adornments.) In the process he had waxed lyrical about their leks. Indeed, the frontispiece illustration for volume 2 of his book *The Malay Archipelago*, engraved by Thomas Wood, shows two hunters concealed in a tree taking aim with their bows at eight male Great Birds of Paradise lekking in the branches above them.

Yet he argued that the astonishing plumage of these paradisiac birds, and the elaborate pattern on a Peacock's tail, were but expressions of surplus energy that was mysteriously granted to one sex by some higher law of the universe. Frankly, when I first understood this I wondered if the old boy had lost his marbles. This sexist theory made no sense as a matter of evolution or indeed as a story of physiology. Or physics, for that matter: a brown feather has just as much 'colour' as a red one, only it is mixed-multicoloured, rather than single-hued. Are we really to suppose that a Peacock has a long and elaborately patterned tail while a Dunnock does not because one has a lot more vigour than the other? Even Rudyard Kipling's Just So stories are rarely this far-fetched.

Perhaps it is no coincidence that *Darwinism* was written when Wallace was becoming ever more deeply immersed in spiritualism and phrenology. He thought an aesthetic sense to be uniquely human and part of a spiritual nature 'conferred on mankind alone by a spiritual act'. In this Wallace was less Victorian than Darwin who, in his drive to stress the continuity of humanity with other species, saw the 'lowest savage' as not greatly more sophisticated than the 'highest ape'. Wallace by contrast had lived among 'savage' tribes in the East Indies and realised their minds were 'very little inferior to [those] of the average member of our learned societies', so he saw the yawning gap between the minds of people – all people – and the minds of animals as needing a special explanation. This led him to reject the idea of animals having an aesthetic sense.

Yet all this was at odds with Wallace's political championing of the rights of women and even his endorsement of the idea that a consistent female preference for gentler men had driven the civilisation of the male sex in human beings. What is more puzzling, though, is that this stance contradicted his entire life's work: in championing natural selection, Wallace had set out to challenge the kind of argument made by natural theology – that things like beauty have no purpose beyond pleasing humankind – by emphasising the selective advantage of every feature. Yet he had now ended up championing the idea that animal beauty was just a neutral outgrowth of male physiological enthusiasm bereft of any utility.

'Subject to periodical fits of gladness'

Darwin had died in 1882 so it was left to others to come to his defence. At least one contemporary observer spotted Wallace's theological inconsistency. Conwy Lloyd Morgan, a student of Thomas Henry Huxley and a pioneering psychologist, in his 1890 book on *Animal Life and Intelligence*, put it this way: 'How, then, does Mr Wallace himself suppose that these secondary sexual characters have arisen? His answer is that "ornament is the natural outcome and direct product of superabundant health and vigour" and is "due to the general laws of growth and development"! At which one rubs one's eyes and looks to the title-page to see that Mr Wallace's name is really there, and not that of Professor Mivart or the Duke of Argyll. For, if the plumage of the Argus Pheasant and the bird of paradise is due to the general laws of growth and development, why not the whole animal? If Darwin's sexual selection is to be thus superseded, why not Messrs Darwin and Wallace's natural selection?'

One other friend and follower of Darwin rushed into print to counter Wallace's argument. George Romanes had become a distinguished biologist while inheriting a large fortune – it was Romanes who had sent Darwin Black Grouse for the table in 1879 – and in the first volume of his book *Darwin, and After Darwin*, he argued that Wallace was missing the point. The ornaments of birds are not just brilliant colours, he pointed out, they are also intricate designs, patterns and structures that were plainly not just the product of exuberant physiology.

> Look, for example, at a Peacock's tail. No doubt it is sufficiently brilliant; but far more remarkable than its brilliancy is its elaborate pattern on the one hand, and its enormous size on the other. There is no conceivable reason why mere *brilliancy of colour*, as an accidental concomitant of general vigour, should have run into so extraordinary, so elaborate, and so beautiful a *design of colours*.

And if it was all about a superabundance of energy, why did such ornaments play a role in courtship at all?

The first volume of Romanes' book, with its chapter on sexual selection, was published in 1893 and was followed by a series of intemperate attacks on Wallace, returned in kind, that were at least partly directed at his socialism. Romanes was at pains to claim that he was only saying what Darwin himself would have said if he were still alive. Indeed he quoted the great man's last words on the subject, read to a meeting of the Zoological Society the day before he died in 1882:

> I may perhaps be here permitted to say that, after having carefully weighed, to the best of my ability, the various arguments which have been advanced against the principle of sexual selection, I remain firmly convinced of its truth.

Within a year of writing his book, Romanes too would be dead.

With most of Darwin's defenders dead, Wallace was left in command of the battlefield. The idea that brightly coloured plumage and melodious song were merely an outpouring of excess vitality was one that attained the status of a settled fact without ever troubling the scorers with actual evidence. In 1889 a treatise on *The Evolution of Sex* by Patrick Geddes and J. Arthur Thomson damned Darwin's theory as 'glaringly anthropomorphic' and firmly endorsed the Wallace view that males were constitutionally built to produce brighter colours 'as the expression of the predominantly katabolic or male sex, and quiet plainness as equally natural to the predominantly anabolic females'. In 1890 the British Museum's spider expert Reginald Pocock insisted that decoration derived from the male's 'high vitality'. In 1891 the travel writer W.H. Hudson waxed lyrical about the extravagant displays of many birds in South America including the lek of the Cock-of-the-Rock, whose stunningly bright orange males gather to dance in a forest glade. But he firmly rejected Darwin's 'laborious' theory in favour of Wallace's idea that male birds are 'subject to periodical fits of gladness' in the spring when 'vitality is at its maximum'. In any case in most species the male 'takes the female he finds', said Hudson unromantically. In 1895 a diplomat turned travel writer and novelist, a close friend of Joseph Conrad named Norman Douglass, wrote a monograph on the topic arguing that 'the harmonious distribution of tints on the feathers of the Argus Pheasant merely continues a principle [of symmetry]' and that it all derived from 'surplus vitality'. In 1903 an influential biologist named Henry Eliot Howard wrote an article on sexual selection in the *Zoologist* in which he argued that Wallace had raised an objection that made Darwin's idea impossible: 'The extremely rigid action of natural selection must render any attempt to select mere ornament utterly nugatory, unless the most ornamented always coincide with the fittest in other respects.' As I shall show in a later chapter this instinctive aversion to sexual selection by female choice among intellectuals inside and outside biology continued well into the twentieth century and remains widespread even today.

The surplus-vitality theory may have been irrational and circular, but it left a legacy of dark obscurity in a corner of science where Darwin had shone a bright light. As the nineteenth century ended, sexual selection in general and female choice in particular were doomed to be largely discarded in the decades to come. As Helena Cronin put it in *The Ant and the Peacock*, whereas Darwin strove to encompass a vast range of colours, feathers, songs and dances within sexual selection, afterwards 'generations of Darwinians strove hard to dismantle that same category' – beginning with Wallace who enjoyed a sort of belated revenge for all the times Darwin had been granted more credit than him. He would prove to be immensely influential in the twentieth century, when his much more limited version of sexual selection, shorn of Darwin's idea of female choice, came to dominate the debate over colours and ornaments in birds: that bright colours are clues to male health. The result was effectively a century without sexual selection as a prominent theory: 'One hundred years in which natural selection was made to account for all the lavish beauty, all the ornamental flourishes that Darwin attributed to mate choice.'

'The most refined beauty may serve as a charm for the female, and for no other purpose'

Darwin and Wallace disagreed about *whether* females choose, but gradually in their debate a key difference had emerged between them as to *what* females would choose if they did choose. For Darwin beauty was an end in itself, desirable but meaningless. For Wallace, it was a means to an end, a signal of quality. Good taste for Darwin, good sense for Wallace, as Cronin put it.

Darwin made the point repeatedly: 'a great number of male animals have been rendered beautiful for beauty's sake'; 'the most refined beauty may serve as a charm for the female, and for no other purpose'; 'that ornament and variety is the sole object, I have myself but little doubt'. Wallace made his case just as often: 'the most vigorous, defiant and mettlesome male' is 'adorned with the finest development of plumage'; 'it is his persistency and his energy which wins the day rather than his beauty'; 'the display of the plumes, like the existence of the plumes themselves, would be the chief external indicator of the maturity and vigour of the male, and would, therefore, be necessarily attractive to the female'.

But each of them left an enigma unsolved. For Darwin it was why the female should prefer beauty at all if it is useless. For Wallace it was why extravagant and arbitrary ornaments, rather than say direct evidence of health and strength, should be such good indicators of fitness. These flaws in their logic stand exposed by the question of whether there is a cost to being gorgeous. Darwin was prepared to concede that conspicuous and exaggerated plumage was not helpful, was maybe even 'slightly injurious', to the bird's survival but not enough to stop the female's taste for beauty influencing the outcome. Going further and admitting that the ornament was a big handicap would put an even bigger question over why females bothered to prefer ornaments at all, to which Darwin had no answer. Likewise Wallace admitted that ornaments must be expensive to grow and handicaps to survival but he thought that as long as this cost was modest it was a virtue, allowing the fitter males to demonstrate their ability to thrive despite the handicaps: the ones that have 'a surplus of strength, vitality, and growth-power, which is able to expend itself in this way without injury'. Both therefore downplayed the costs of being gorgeous.

Sorry, but this won't wash. The birds I am watching right now have not just grown, but preened and cleaned all winter feathers of the

smartest pure colours: blue, black and bright white. They have wasted
hours every day defending a tiny territory on a lek that has no food in
it. As the spring has advanced they have spent more and more time
roo-kooing and sneezing incessantly for hours on end, constantly on
the move and expending energy. They may even have invested as many
calories in display as the females will soon invest in laying and incubating eggs. They have returned to the same spot day after day risking
predator attack. They will have lost weight, probably been injured in
fights and possibly become sick. No way is the cost of doing all this
small. If Darwin and Wallace cannot explain why such exceedingly
costly behaviour evolves, they cannot explain the lek at all.

Let me take another careful look at a fourth female now scoping the
males on the lek. What is it that will decide her choice of a mate? Is she
really able to discriminate a slight superiority in the frequency or intensity of the dance of the central bird? Surely the marginal value his extra
beauty – or his extra fitness – brings is infinitesimal either way. So
maybe it is the fact that he is in the centre of the lek that counts. But if
so, why? What on earth could it matter that the male in the middle is
Mr Right? It's not as if he is king of some castle: his well-trodden patch
of moss and grass is just like any other. So here's a curious thought.
Maybe what is going on is that all the females are watching each other
and making sure they choose the same male, not because he's better or
prettier or more central but just because they want to do the same
thing as each other.

The idea of females copying each other's choice of mate first occurred
to scientists studying Sage Grouse leks in the American west. A man
named Haven Wiley watched three Sage Grouse leks in three consecu-
tive springs in the late 1960s and his descriptions of how females
behave there are similar to mine. His male birds strutted incessantly
from well before dawn, fought occasionally and entertained highly
discriminating females that arrived around sunrise by air, then walked
towards the centre of the lek, chose central males and solicited sex from

them. Wiley's leks were huge, however, with up to 260 males at one, and at the height of the mating season in early April, the females formed tight packs in the central territories.

The skew in mating success in Sage Grouse is even more extreme than in Black Grouse. Wiley once saw a male perform 34 matings in a single morning, and in a later study a single male was seen to make 169 in one season.

How did females pick the male they mated with? Wiley could detect no greater vigour in the display of the successful male. Indeed the reverse: 'In 1967 the most successful breeder also showed signs of fatigue on mornings when females were numerous. Late in the morning his matings often became noticeably slower, and he was slower to resume strutting.' After twenty matings or more, it's not surprising! Nor could Wiley detect anything about the vegetation or topography of the lek centre that seemed special. The mating centres – on the

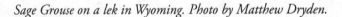

Sage Grouse on a lek in Wyoming. Photo by Matthew Dryden.

biggest lek there was more than one – were placed right in the dense middle of the males but not necessarily at an exact geometric centre.

Instead Wiley wondered whether the females used cues from other females to locate the mating centre. He observed that as females moved towards the lek centre they were more likely to pause when they encountered a group of other females. By the time they reached the mating male they were in quite a crowd. Like crystals precipitating from solution, they aggregated naturally. Was it possible that the main attraction of a male was his attractiveness itself to other females, nothing more? Wiley placed a few taxidermist's stuffed females elsewhere in the lek to see if he could create an artificial hot spot but these failed to attract females as he had hoped.

The zoologist David Lack had tried the same trick with Black Grouse in Scotland in 1938, placing a dead, stuffed female and two stuffed males in the territories of three different males. One of the male dummies was promptly decapitated by the nearest cock, while each of the three males offered the female dummy copulated enthusiastically with her (it?). The first two were interrupted after copulating 'a few times'. The third was allowed to continue for forty-five minutes 'in which time it copulated as many as 56 times and an area of 3 by 2 inches on the mount was soaked with seminal fluid'. Poor blighter. Lack did not record whether the dummy served to enhance the reputation of the male in whose territory it was placed, as the copying theory would imply.

You might think Lack had gone a little far, ethically speaking, with this experiment, but a Dutch and an American scientist said, 'Hold my beer!' J.P. Kruijt and Jerry Hogan repeated the very same experiment in 1962 in the Netherlands, setting up their stuffed female beneath a cloth, which could be pulled away with a string at a suitable moment, and they broke Lack's record. Their stuffed female was subjected to multiple half-minute copulation attempts on different days by eight different males, which continued in each case for never less than half an hour and in one case for two hours. In between mountings, the frustrated males kept returning to the crouch display in an increasingly desperate attempt to get a bit of cooperation from the dummy. Poor birds.

The sexual deception of long-suffering Black Grouse continued, though this time with a bit more of a result. In the 1990s Rauno Alatalo and his colleagues repeated the experiment with stuffed female Black Grouse in both Finland and Sweden. These scientists had already observed that for a particular male, copulations came in streaks more than would be expected by chance. That is, once a male had success-fully mated with one female he was more likely to mate with a second and third. So, on seven occasions over four years the scientists placed four dummy (stuffed) female Black Grouse in the territories of males that would not otherwise be likely to mate. The dummies were put there before the lekking began in the pre-dawn darkness and placed on the ground so that males could both display to them and, if they wished, copulate with them. In one of the four years when they did this experiment, the scientists placed the dummies on sticks, as if perched on a small tree overlooking the lek as females like to do. And finally as a control they also placed three plastic duck decoys in males' territories to see if they were courted. This was to check that the birds did indeed discriminate the grouse dummies from other irrelevant objects.

The results of these experiments – apart, presumably, from creating traumatic psychological issues in a number of male Black Grouse – were revealing. Once again the males displayed at the dummies and mounted them repeatedly, trying for many minutes to inseminate them, as in the Scottish and the Dutch experiments. But the Finns observed something new. On mornings when a particular male was provided with dummies on the ground, other females spent more time in his territory than they had the previous day or would the following day. This was not the case for the dummies on sticks or for the duck decoys. But one morning a Goshawk plunged into the lek and tried to carry off a duck decoy, suggesting that the ducks were seen as real by birds.

In 1987 and 1988 in an experimental facility near Laramie, Wyoming, my old friend and colleague Professor Mark Boyce used a captive flock of Sage Grouse to test whether copying happened and what criteria the females were using to select a mate. Each day a female was placed in a six-sided cage with a male caged on each of the six sides. The positions of the males were randomised for each trial, changing their location every few days. To see if copying occurred, three of the six males were sometimes accompanied by a female in their pens. Females tended to position themselves close to the spot in the arena where the most strutting was happening, meaning that both the individual display rate of a male and the rate of display of his neighbours seemed to attract them. (One young male displayed enthusiastically but idiosyncratically. At the climax of his strut, when he was supposed to push his olive-green, boob-like air sacs out through his white plumage three times, he kept doing it four times. The females were unimpressed.)

The females were not especially drawn to males accompanied by females. And when four different females were given precisely the same arrangement of males, one after the other, they tended to choose the same male, even though they had not witnessed the others' choice. In other words, females may be independently choosing the same male on a lek without any need to copy each other. They are not necessarily interested in males that have already charmed other females, to borrow Darwin's vocabulary, just because of that.

Of course, the experimental set-up in Laramie might not have been ideal for deciding this matter and copying mate choice may well still happen in the wild as suggested by the model experiments in Finland. In any case, preferring the same male as other females for whatever reason is the opposite of most female birds in other species, which apparently prefer an unmated male so that they can hope to get his input into rearing young – which is why they end up in pairs. There may be some definite benefit to a female in choosing the same male as other females in Black and Sage Grouse. Or to put it in evolutionary terms, females that choose the same male on a lek as each other must leave more offspring. What that benefit might be will have to wait for the next chapter, but note here how dangerous it might be not to follow the fashion under such a system. If a female were to decide to go her own way and ignore the crowd, she might end up breeding sons that were not so attractive to the opposite sex.

4.

Runaway Fashion

BOYET: Be now as prodigal of all dear grace
As nature was in making graces dear
When she did starve the general world besides
And prodigally gave them all to you.
PRINCESS: Good Lord Boyet, my beauty, though but mean,
Needs not the painted flourish of your praise.
Beauty is bought by judgment of the eye.

WILLIAM SHAKESPEARE, *Love's Labour's Lost*

6 a.m., April, the Pennine hills

The sun catches a glacial erratic, a rock that was floated here by the ice sheets and dropped when they melted eleven thousand years ago, on a slope half a mile to the south-west of the lek. I know that rock; it's marked with strange patterns. These are cup-and-ring marks, left by either Neolithic or Bronze Age people up to four thousand years ago for reasons we still cannot divine. Deeply engraved into hard rocks, the saucer-shaped hollows and concentric grooves are the work of many hands over many years. Their like can be found throughout the northern and western parts of the island of Britain as well as elsewhere in Europe. As to their meaning or purpose, theories abound, but none persuades. Perhaps they marked the territorial boundaries of tribes. Perhaps they were used to measure out small quantities of gold or grain as payment. Perhaps they recorded the seasons. Perhaps they were maps of the stars. Perhaps they were of ritual use, to hold the blood of a sacrificed goat. Or perhaps they were just random graffiti and we're wrong to over-interpret them.

Indeed the erratic rock itself is testament to the randomness of causes in nature. In one of his forays into science, the great literary polymath Johann Wolfgang von Goethe was the first person to explain erratics as having been deposited by ice. Goethe's various fragments of essays about ice, written in the 1820s, including one called 'Erratische Blocke', grappled with the mystery of huge lumps of granite found lying on the North German plain. In his novel *Wilhelm Meister's Journeyman Years* (1821), Goethe fancied that during 'a period of grim cold … the transport of big blocks was made possible by floating drift ice'. In 1810 Leopold von Buch had proved that these erratic blocks came from Scandinavia, but he adamantly dismissed the ice theory. Goethe stuck to it, though it was floating ice, not land ice, he had in mind: 'Big ice rafts carry granite blocks in the Baltic,' he wrote in his fragment. Later at Lac Léman he saw proof that he was right. 'The glaciers travel through the valleys to the edge of the lake carrying the granite blocks loosed from above, as still happens today.' But as so often in science, dogma prevailed for decades, in the form of von Buch's insistence that the German plain had never been under ice. Geologists Louis Agassiz and Charles Lyell eventually proved Goethe right long after he was dead. The position of these erratic blocks is down to accident, not purpose or design.

As I watch the Black Grouse roo-kooing in front of me, it occurs to me that there is a parallel here. Are we prone to over-interpreting the purpose of growing a long tail, wearing bold colours and strutting energetically in every dawn? Am I trying too hard to find purpose in every detail? I suspect we may need to be more open to the possibility that sexual selection is a phenomenon that generates arbitrary ornaments, devoid of practical purpose and useful meaning, just as cup-and-ring rocks may have had a general ritual purpose but that the detail of the decoration is not full of meaning any more than abstract art is today. Hence the sheer randomness of a lyre-shaped tail or a white bum or a roo-kooing that sounds like the idling engine of a UFO being a tool of seduction, like the sheer randomness of a lump of Swedish granite on the North German plain or the sheer randomness of human-made patterns on the rocks. And for Black Grouse tails there is a fascinating theory that suggests just this and – perhaps not coincidentally – is never very popular for long with scientists because it leaves them little to test. It's called runaway.

When Darwin died, you recall, his embarrassing idea of female choice as a key driver of evolution mostly died with him. Wallace's version of sexual selection, that display was just another way of demonstrating vigour and perhaps fitness to survive, survived, but any notion of selective female choice faded. Males decided among themselves who would mate, through battle. Worse, the crusty alpha males that dominated biology went even further and insisted that if anything the very purpose of display and ornament was to intimidate other males or to overcome the coyness of females. The females were mere bystanders (or victims), meekly accepting the outcome of a competition to flaunt the fittest genes.

Thomas Hunt Morgan, whose series of discoveries laid the foundation of genetic studies in fruit flies, was especially scathing about female choice in a book about evolution published in 1903. Describing the 'heaping up the ornaments on one side and the appreciation of these ornaments on the other' as a 'fiction' that 'meets with fatal objections at every turn', he mocked the notion that 'women have caused the beard of man to develop by selecting the best bearded individuals' while men select the women 'that have the least amount of beard'.

(There's the obsession with whiskers again!) Fourteen years later, in 1917, the Scottish biologist and mathematician Sir D'Arcy Thompson published a highly original take on evolution in the form of a book called *On Growth and Form*, but he dismissed female choice theory curtly: 'The jewelled splendour of the Peacock [is] ascribed to vanity in one sex and wantonness in the other.'

'The most vigorous, defiant, and mettlesome'

The problem had not gone away, though, and some theory was needed to explain why displaying colourful feathers made sense even as tools of intimidation. Throughout the birds, as Darwin had exhaustively catalogued, there were myriad examples of male creatures sporting adornments of no practical use, let alone as weapons: usually with striking and bold colours and always with elaborate and elegant dancing or singing to display them in the presence of females. The faintly ridiculous idea that these ornaments got there by being bred into males by selective females, as if they were bantam breeds, was frankly as good a candidate as any. And it was implicit in Wallace's concession that the purpose of male competition might be to give the nuptial spoils to the fittest males: that was still a sort of choice.

It was in Goethe's Germany that signs of life stirred within the slumbering body of Darwin's idea. In the 1880s August Weismann gave a positive assessment of Darwin's theory of sexual selection in his lectures at the University of Freiburg, later published in his book *The Evolution Theory*. He dismissed Wallace's vigorous-male alternative, saying, 'It is not easy to see why a more active metabolism should be necessary for the production of strikingly bright colours than for that of a dark or protective colour.' He called mate choice 'a much more powerful factor in transformation than we should at first be inclined to believe' and made the case that 'females do not behave as dispassionate judges, but as excitable persons which fall to the lot of the male who is able to excite them most strongly'.

Darwin's most prominent German disciple and friend, Ernst Haeckel, had a disciple and friend, too, who became a novelist and nature writer: Wilhelm Bölsche. In 1898 Bölsche wrote a bestseller entitled *The Love Life in Nature: Or the Story of the Evolution of Love*, which embraced the theory of sexual selection with salacious enthusiasm. As the philosopher David Rothenberg put it, Bölsche put the sex back into sexual selection, writing that: 'Love as a spiritual value of the very highest kind ever anew strives upward on proud eagle's pinions high into this same blue above the chemical mysteries of the sexual act [causing an] infinite golden wave that floods everything which love has even touched.' He was quite prepared to grant such orgasmic ecstasy to birds of paradise and bowerbirds, writing of their bewitchment during the act of love, their intoxication and transport to 'a liberated aesthetic inner life'. Such anthropomorphic soft porn may have made Bölsche's writings bestsellers, Rothenberg comments, but probably did not endear the theory of sexual selection to more meticulous, empirical minds.

Two years earlier the German philosopher and psychologist Karl Groos, at the University of Giessen, had authored a widely admired treatise on 'the play of animals', first published in 1896 in German, in which he devoted a chapter to sexual selection. He accepted Wallace's misleading claim that females do not seem to be deliberately choosy, let alone in a way that would consistently produce the exaggeration of particular features of males. But he could not swallow the logic of surplus vitality: 'I for one can not quite conceive how such developments as, for instance, a Peacock's tail, can be derived from beginnings so insignificant, simply by a superabundance of energy.' And he seized on Wallace's suggestion that brilliant colour could be an indication of robust health, allowing females inadvertently to pick or be picked by 'the most vigorous, defiant, and mettlesome' males. Surely that meant

Darwin might have a point: 'Wallace overturns his whole argument, for if it is once admitted that the female chooses the strongest male, the chief point of the Darwinian theory is conceded. Whether her preference is for strength and courage or for beauty is of little consequence; the important thing is that a choice is made.'

Groos thus rescued Darwin from Wallace's mistaken censure but he did so while still insisting that female choice was an involuntary phenomenon, driven by the impulse that more brilliant males were better able to overcome female 'coyness' and reluctance to mate. To do otherwise would be to shock the sensibilities of the age. 'The probability is that seldom or never does the female exert any choice,' Groos wrote. Females were the 'hunted hare', not the prize giver, he said, true to his time. But he came close to conceding even on this point. After all, he added in a direct echo of Darwin, a soldier in his handsome uniform is more attractive than the same man in his working blouse.

Groos's 'unconscious choosing' dissolved the greatest objection of Wallace and his followers to female choice theory, namely that animals could not be expected to exercise sophisticated aesthetic judgements. So a procession of incredulous professors had insisted. The Polish ornithologist Jean Stolzmann, for example, told the Zoological Society of London in 1885 that 'at first sight, it is difficult for us to admit in the female birds the presence of an aesthetic taste so strongly developed as the Darwinian signal [requires]'. Patrick Geddes and J. Arthur Thomson, authors of *The Evolution of Sex* (1889), agreed: 'It seems difficult to credit birds or butterflies with a degree of æsthetic development exhibited by no human being without special æsthetic acuteness and special training.' Yet this objection made little sense. A person can catch a ball without having been taught differential equations or enjoy a glass of champagne without an education in wine; so a bird can appreciate beauty without having been on a university course in aesthetics. As Lloyd Morgan put it, 'The hen selects that mate which by his song or otherwise excites in greatest degree the mating impulse; and there is no more need to suppose the existence of an aesthetic standard in this case than there is to hypothecate a gustatory ideal in the case of a chick that eats a juicy worm.'

Yet, as Helena Cronin pointed out, Darwin had very clearly not taken this route around the objection, eschewing the chance to put female taste for beauty in the same instinctive category as the ability to choose nutritious food. He had written in *Descent*: 'I fully admit that it is an astonishing fact that the females of many birds and some mammals should be endowed with sufficient taste for what has apparently been effected through sexual selection; and this is even more astonishing in the case of reptiles, fish, and insects.' He really did want female birds to be capable of sophisticated aesthetic judgements. Why? Cronin thinks it is clear from his notebooks that it was part of his

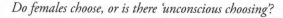

Do females choose, or is there 'unconscious choosing'?

perpetual campaign of showing continuity between people and other animals: if he could show that a taste for the beautiful was not confined to human beings, then another example of human exceptionalism would fall. And sexual display was the place to look, Burke having persuaded him that the human taste for beauty was erotic in origin. 'Beauty ... cuts the knot,' he wrote at one point. For Wallace, an appreciation of beauty marked people out from other animals but for Darwin it brought them closer together.

'The enticing idea of the all-puissance of natural selection'

In the early years of the twentieth century an ornithologist called Frank Finn, based in Calcutta where he was deputy superintendent of the Indian Museum, did at last do the obvious experiment, albeit only once. He took a small female waxbill finch, called the Red Munia, and placed it in the middle section of a three-compartment cage. In each of the other compartments he placed a male of the same species. The male Red Munia when adult develops a bright red bill and scarlet plumage spotted with white. Finn takes up the tale: 'I submitted to the hen first of all two male birds, one of a coppery and the other of a rich scarlet tint.' The female took to roosting next to the scarlet male. The duller male promptly died, of grief Finn assumed: 'A warning to future experimenters to remove the rejected suitor as early as possible.' He repeated the experiment with two new males with the same result, though fortunately no death: the brighter bird was clearly preferred. He then tried a new female but she expressed no preference. It's hardly the definitive experiment, but at least he tried.

Finn and his co-author Douglas Dewar were no fans of Wallace and his followers, who they thought were too obsessed with natural selection. 'To such an extent does the enticing idea of the all-puissance of natural selection dominate the minds of scientific men that but few of them have paid any attention to the question of sexual selection.' But nor did Finn and Dewar much rate Darwin: 'It is argued, with some show of reason, that it is absurd to credit birds with æsthetic tastes equal, if not superior, to those of the most refined and civilised of human beings.' They were part of the early twentieth-century vogue for believing that the discovery of genetics did away with the need for Darwinian selection altogether (their book *The Making of Species* was published in 1909): 'Our view, then, is that evolution proceeds by mutations, which may be large or small.'

'Their sons are finer and get more grand-chicks'

A revival of Darwin's female choice awaited a rationale for it. And one was now in the offing. Selous had made a pointed dig at the book-writing sages who pontificated from their libraries about the lack of female choice in nature without ever going into the field and looking. 'I believe that, as denial from the chair is replaced, or supplemented, by evidence from the field, the views of that great naturalist and reasoner [Darwin] will be triumphantly and often most strikingly vindicated.' One chair-bound scientist who took notice of Selous's words was Ronald Fisher, a mathematician, biologist and pioneer of modern statistics. In a long career Fisher laid the foundations of statistical analysis, helped integrate genetics with evolution, co-founded the field of population genetics and heavily influenced the science of crop breeding. However, he also gave scientific support for some of the horrible excesses of eugenics.

In a letter in the 1930s, Fisher wrote that: 'Selous' observations on the Ruff, where he has seen the hen passing with perfect self composure among the crowd of males, who await, but cannot hurry, her choice, provide a perfect ecological framework for this runaway type of selection. The hens choose the fashion of their sons' ornaments.' Darwin's vindication at Selous's leks had not gone entirely unnoticed. (Selous's Ruff observations are described in chapter 10.)

Fisher's new idea lurked inside that word 'fashion'. He had glimpsed a possibility of arbitrary exaggeration or what he called 'runaway type of selection'. To understand how he came to this insight, it is useful to trace the history. In 1915, while a young schoolteacher, he wrote a short paper entitled 'The Evolution of Sexual Preference' in the *Eugenics Review* that made a simple point about sexual selection. The issue was not whether females choose, so much as why they choose, he said. While Wallace's criticisms of whether females choose were 'weak', the question Darwin had been allowed by his critics to duck was why females should choose at all, if they do, let alone in a consistent way shared with other females in the same species. 'Why have females this taste? Of what use is it to the species that they should select this seemingly useless ornament?' Darwin had implied that female birds like beautiful plumage just because they have a taste for the beautiful. This circular reasoning – rather than the lack of evidence for females being choosy – was the weakest link in his argument, Fisher thought. One anonymous reviewer of Darwin's *Descent* had spotted this at the time, writing that female choice was 'a cause, which will seem to most men more needful of explanation and more worthy of it, than the effect itself'; but on the whole he had got away with this omission.

Having posed the question, Fisher's 1915 article did not answer it, wandering off into some rather vapid speculation about human mating preferences and standards of beauty that reads as ill-judged today. Eugenics was then very much in vogue all across the political spectrum and Fisher was an enthusiastic supporter of the idea that people should be encouraged, cajoled, perhaps even forced into selecting their mates carefully for the good of the human race – and not breeding at all, or being allowed to breed, if they had various kinds of defects. The impracticality of this fashionable proposition, and its encouragement to cruel and even genocidal policies, had only begun to be glimpsed by a few critics. Fisher was not one of those critics and his enthusiasm for eugenics even continued after the Second World War, so there is much about his writing that sticks in the modern craw, even while his statistical innovations continue to be widely used and his pioneering work on genetics underpins modern evolutionary biology.

But here's the intriguing feature of Fisher's particular version of sexual selection by female choice when he eventually spelled it out in 1930: it's effectively anti-eugenic. Or at least it suggests that another force will take over from any eugenic habit of choosing the fittest mate, and crowd it out. He argued that females could get into the habit of choosing arbitrary ornaments whether or not they indicated health and fitness. Fisher now provided an answer to his question of why females should develop strong preferences in his book *The Genetical Theory of Natural Selection*. That book was overall a landmark in the history of biology for it showed that Gregor Mendel's discovery of the genetic mechanism of heredity was not just compatible with Darwin's natural selection – contradicting the conventional wisdom of the time – but essential to it. The understanding that inheritance came in the form of discrete particles, rather than blending fluids, that could pass through a generation unchanged or unexpressed to emerge in the next one, proved vital in showing that Darwinian selection would not be undermined by dilution at every mating.

But the book is just as famous for reviving Darwin's theory of female choice with an ingenious twist. Fisher saw that in order to have descendants, an animal must not just have lots of offspring but must have offspring that themselves breed successfully. There is no point in having ten children if they all die young. This means choosing a healthy mate if possible or providing good child care. So far, so eugenic. But it means more than that: one measure of the quality of the offspring will be their ability to attract a mate. There is no point – in Darwinian terms – in having lots of young if none of them gets to mate because other birds find them unattractive. Especially in a polygamous species with elaborate display, like Black Grouse, a female who mates with a dull or clumsy male risks producing dull or clumsy sons that remain unmated. As Fisher put it in a letter to Darwin's grandson, 'My theory is that tasteful hens don't rear more chicks, but their sons are finer and get more grand-chicks.' So it makes sense for a hen to follow the fashion, choose the most glamorous male and thereby enhance her chance of winning the male genetic jackpot in the next generation.

A female Black Grouse that chooses the central male on the lek (the one who may make a score of matings in one season) is hoping –

Choosing a glamorous male should mean having glamorous sons.

unconsciously of course – to see her offspring occupy the centre of the lek themselves in a few years' time. Perhaps that is why she jealously tries to chase away other females, as Selous observed and I witnessed. Fisher did not use the phrase 'sexy son hypothesis' but that is what this idea has come to be called.

In a choosy species, it follows that the stronger the female's preference for particular males, the more successful she will be at leaving descendants, so long as other females admire the same kind of male. Therefore the preference will get stronger and more correlated over time, as mutations that intensify the selectivity of the females are favoured. This puts the males to the test. Females are getting more choosy so males have to become more persuasive, which means larger ornaments and more vigorous displays. Fisher had spotted that there would be an arms race between the two sexes and that both the preference and the ornaments would co-evolve in lockstep together.

Notice I wrote 'so long as other females admire the same kind of male'. Now suddenly copying others' mate choices makes sense. Could it be that every female is subconsciously looking around the lek and checking that she's making the same choice as most of the others? Hence Wiley's finding that Sage Grouse hens move in a flock and seem to be attracted to males that have a gaggle of females around them already. Hence Alatalo's finding that he could enhance the attractiveness of a male Black Grouse on a lek by placing stuffed female dummies in his territory. Hence too a common scene filmed at Capercaillie leks. Go online and look up videos of displaying Capercaillies in Scandinavia and you will soon find footage of a displaying cock surrounded by half a dozen adoring hens, whose eagerness to mate is plain to see: each keeps squatting in front of him when he faces towards her. He keeps ignoring these invitations and continues strutting his display in apparently splendid self-absorption. Why? Perhaps if females are copying each other's choice, one way for the male to maximise his matings is to keep the females waiting while more and more are attracted to gather around him. The tendency of a male to seem more attractive to females because he is admired by other females is not unknown even in human beings, remember: it has been called the 'wedding-ring effect'. But in a lekking bird, subject to the sexy son effect, it is almost bound to be a strong factor. Fisher and mate copying go together.

Crucially, Fisher realised, this sexy son effect will continue even if the ornaments themselves are giving no reliable information about the quality of the male in other respects. Once female preference is established, the females are slaves to fashion. They dare not choose differently lest they have unsexy sons. A Peacock that grows no long train, and is thereby saved the physiological effort, could – probably would – become heavier, healthier and more energetic. A Black Grouse male that did no displaying, conserving his energy every morning, would remain healthier and heavier. All the energy that would have gone into the display is redirected into slightly bigger muscles, a slightly larger brain, a slightly stronger immune system. But it doesn't help him. The Peacock's pitiful attempts at tail-less display, or the Black Grouse's idle refusal to roo-koo and sneeze, will not impress the females. (See what I mean about anti-eugenic?) As ornithologist Richard Prum argues, 'The very existence of mate choice would unhinge the display trait from its original honest, quality information by creating a new, unpredictable, aesthetically driven evolutionary force: sexual attraction to the trait itself … Desire for beauty will undermine the desire for truth.'

Thus the process will continue to enlarge the ornaments of birds even when they become handicaps that diminish the survival chances of the males. Darwin and Wallace were both reluctant to emphasise the huge costs to males of growing and displaying large ornaments, but Fisher shows that males of a highly polygamous species might experience a huge cost and still be better off bearing it. So long as the disadvantage is more than counterbalanced by the advantage in sexual selection, he wrote, then further development will proceed. In mathematical terms a Peacock will grow a train that halves its chances of surviving if that tail more than doubles its chances of mating before it dies. Reproduction of the sexiest trumps survival of the fittest.

And boy was he right about the cost! Black Spot, the top cock on the lek this year, will probably not be back next year. In chapter 5 I will show data that suggests that each male Black Grouse gets one shot at being the top cock, one season of maximum effort, at age three, four, five or even six, when he has worked his way towards the centre of the lek and now throws everything at persuading the females he is the right one to choose, attending every day, displaying incessantly. In the process he exhausts and sickens himself.

Fisher suggested that extreme plumage development in the male, and preference for extreme plumage in the female, 'must thus advance together' and so long as nothing else stops them, this will happen at 'ever-increasing speed'. There is therefore, he said, 'the potentiality of a runaway process, which, however small the beginnings from which it arose, must, unless checked, produce great effects and in the later stages with great rapidity'. At last, here was a theoretical explanation of orna-ments as extravagant and as useless and as harmful to survival as Peacocks' trains.

This is an extraordinary idea. Let's translate it into a human analogy and see how it feels. In a Bronze Age settlement, young men are obsessed with carving cup-and-ring markings on rocks. Nobody can recall how it started or why, or what purpose the carvings originally served. But young women will not marry a man until he has carved a good cup-and-ring mark on a rock. Nobody can recall why they have this preference, but it has become an obsession. A woman who chooses a strong and intelligent man who has carved no marks faces evolution-ary oblivion because she risks having sons that fail to marry, because they fail to learn to carve. So none dares buck the trend.

All right, the analogy does not work well, I admit, not only because I have no evidence that cups and rings were carved by men trying to seduce women (though it's as good a theory as any other) but also because people are conscious, because mating preferences in human beings are far more complicated, because people are mostly monogamous, and so forth. Also because people do not belong to a species with extravagant secondary sexual characteristics, unless you call the human mind an extravagant sexual ornament – an idea I shall flirt with in chapter 12. But the tyrannical nature of the process, imposed by a sort of despotic fashion, is made clear by my analogy.

Of course, the runaway process that Fisher described is unconscious and has no foresight. It simply emerges as a byproduct of female choosiness, however slight that choosiness originally is. All he is saying is that in species where females are choosing based on plumage, and where males are highly polygamous, that choosiness will accelerate, and so will the chosen male feature to an absurd and arbitrary conclusion, or until checked by some limiting factor. Hence it is futile to look for meaning in the train of a Peacock or the roo-kooing of a Black Grouse. Its beauty is its own reproductive reward: it works as a seduction device because it works as a seduction device. Taste for beauty causes beauty which causes taste for beauty. Vindication for Darwin and his insistence on 'beauty for its own sake' at last!

'Through the abstract space of possible ornament designs'

One consequence of Fisher's logic is that species like Peacocks may have, in the past, suffered sudden and rapid bursts of sexual selection that perhaps did not last long. The train may have grown to its vast modern size in a burst of runaway change lasting just a few hundred generations. Catching the runaway process in the act may therefore be tricky, a point I shall return to.

A further, rather beautiful consequence of Fisher's idea is that it can explain why the ornaments of birds are so arbitrary. In one species the tail is exaggerated, in another the wing, in a third the neck wattle, in a fourth the crest. In cardinals the male is red, in Black Grouse black, in bluebirds blue, in orioles yellow. In Nightingales the male sings, in Argus Pheasants he dances, in Black Grouse he does both and communally at a lek. Fisher would argue that sexual selection by runaway female choice has latched on to some feature of male plumage that just happened to help in seducing females and exaggerated it to spectacular effect along with an ever more exaggerated female obsession with that feature. It's as meaningless as a good painting or a fine symphony – meaningless but wonderful. In the grouse there are half a dozen species that lek (depending on how you define a species and a lek), but each employs a completely different kind of dance, has completely different colours, grows completely different elongated feathers and inflates completely different throat sacs. As I shall show later, birds of paradise are even more diverse in their plumage and displays.

In the 1990s at Stanford University Geoffrey Miller and Peter Todd ran simulations of runaway sexual selection and found that tiny differences in starting conditions could drive virtual creatures in wildly different directions with wildly different ornaments and wildly different preferences. 'If you went out for a coffee while running a simulation and came back ten minutes later, the population would usually have moved where you least expected it – not through the physical space of its simulated habitat, but through the abstract space of possible ornament designs,' reported Miller in his book *The Mating Mind*.

Fisher pointed out that the runaway process would not happen to male weapons used for fighting with other males. Having excessively vast antlers is not much use to a deer unless it really is strong enough to win a fight. For a Peacock, which can grow a vast, 'dishonest' train, there is no such thing as too big because the bigger it is, the more likely it is that the male will win the mating jackpot. So ornaments would only grow to absurd and useless size if chosen by female choice, rather than by male competition.

'The cock is only fine because the hen is artistic'

In 1930, when his book was published, Ronald Fisher exchanged letters with a physicist friend who happened to be Charles Darwin's grandson: Charles Galton Darwin. This namesake did not share his grandfather's confidence in female choice let alone Fisher's extension of the idea. 'As to the runaway process of sexual selection I am still unconvinced,' wrote Darwin. For a runaway process there need to be two causes reinforcing each other, he argued: 'You say fine cock is one and artistic hen the other. I say that the cock is only fine *because* the hen is artistic, and I claim that is only a single effect.' Fisher replied that he would have another shot at explaining runaway selection. It would be two years before he did so, writing to Darwin on 25 October 1932 with what turned out to be the first mathematical model of the phenomenon of runaway selection. This correspondence was only discovered in 1999. Till then it had been assumed that Fisher had never done the mathematics on his idea to vindicate his hunch that it worked.

'Take x for cock beauty and y for hen taste,' Fisher began, and proceeded to derive a formula for the rate at which each would drive accelerating change in the other given certain assumptions about the variation in male beauty and the strength of female discrimination. The mathematical modelling proved to Fisher's satisfaction that his idea was indeed highly plausible, including the accelerating rate of change in the ornament. C.G. Darwin proved Wallace-like in his obstinacy and now replied that he did not see how this was different from a natural selection process affecting only one sex such as milk yield in cows. Fisher responded that the exponential acceleration in sexual selection came from the 'rate of change in hen taste being proportional to the absolute average degree of taste', a feature that does not apply to other sex-linked traits. Milk yield may be transmitted through a bull but 'the intensity of selection in favour of higher milk yield is not determined by the average milk yield'. What he was saying here was that whereas in most features of animals the more extreme they became, the less they would be favoured to become more extreme – diminishing returns – here it was the opposite: the more selective the females got, the more selection favoured even stronger selectivity – increasing returns. Jonathan Henshaw and Adam Jones of the University of Idaho analysed Fisher's model in 2019 and concluded that it contained all the right ingredients but would not have quite worked.

Here's a strange coincidence. What Fisher is suggesting is effectively a 'chain reaction': cause accelerates effect which accelerates cause. That phrase was about to become much better known in a very different context. Less than a year after Fisher sent the proof to Darwin, on 12 September 1933, the Hungarian physicist Leo Szilard, while waiting for a traffic light to change at the junction of Southampton Row and Russell Square in London, suddenly saw in his mind that under certain conditions nuclear fission, by emitting neutrons that cause more fission that emits more neutrons, could cause a chain reaction and hence a massive explosion. The atom bomb became a terrifying possibility for the first time.

Fisher never published his mathematical proof of the concept of runaway selection. Why not? He was a prolific publisher of mathematical proofs in genetics generally and he recognised this as a unique contribution to a unique Darwinian idea. Perhaps he realised it undermined his eugenic ideas. If you told people that mate choice had a habit of fixing on some arbitrary, meaningless and unstable definition of what was attractive, you could hardly also urge them that it was their duty to choose healthy mates for the good of the species. I'm not so sure. But the fact that his mathematical proof remained hidden in private correspondence certainly hindered the runaway theory's chances of being taken seriously.

'Darwin's original contention will not hold'

Fisher, like Darwin (the evolutionist not the physicist), was to be largely ignored. From the 1920s to the 1970s biologists who studied sexual behaviour in birds were monotonously insistent that display and ornament were the result of male competition or something else, but almost never female choice. In particular, Julian Huxley's influence on the sexual selection debate was to be long-lasting and in my view almost wholly malign.

Huxley, the Oxford-educated grandson of Darwin's 'bulldog' Thomas Henry Huxley, became more of a scientific administrator and populariser than practitioner, ending his career as the first head of UNESCO. Unlike the reclusive Selous, who he knew, he was the sort of accomplished networker who often ends up at the top of the scientific tree. As a student, Huxley was a keen bird watcher and emulated Selous in paying careful attention to bird behaviour. On holiday in Wales he watched Redshanks courting, then Oystercatchers in Holland and then Great Crested Grebes on a lake in Hertfordshire. In 1921 it was Red-throated Divers on the Arctic island of Spitsbergen. All these birds formed long-lasting monogamous pairs in the breeding season yet indulged in ritualised displays. This turned Huxley into a dedicated sceptic of the idea of female choice. He went so far as to claim in his paper on divers in 1923 that 'the majority of the "courtship" actions which are to be found described in the literature occur in the post-mating period, and therefore cannot be operative in any true form of sexual selection as imagined by Darwin' – which would be news to Peafowl, Ruffs, Black Grouse and many other species.

Yet it is hard to pin Huxley down on the topic. In 1927 he co-authored a book on animal biology with J.B.S. Haldane in which they passingly mentioned that in the case of male ornaments selection is 'exercised by the mind of the female'. That same year Huxley actually wrote the introduction for Edmund Selous's book *Realities of Bird Life*, which reprinted the latter's diary journals on various species, including Ruffs and Black Grouse. Huxley acknowledged his debt to the older man as a pioneer of careful and detailed observation of bird courtship and conceded that Selous had shown Darwin's principle of sexual selection 'definitely held good for' the Ruff, though he seems not to have read the passages on Black Grouse carefully if at all. In 1930 he wrote a book on bird watching and agreed that 'for Ruffs at least the essence of the Darwinian hypothesis seems proved'.

So at this stage he was prepared to concede the possibility that female choice could cause male display, but felt it to be a minor curiosity in a few highly polygamous species with no wide-ranging importance in the theory of evolution. In 1929 (while his brother Aldous was writing *Brave New World*) Julian co-authored an 895-page treatise on *The Science of Life* with H.G. Wells, in which he curtly dismissed sexual selection, writing that display 'prompts to the act of mating rather than to the choice of marriage partner'.

By the 1940s Huxley had come to be widely admired as a leading theorist of evolution who had combined and summarised the 'modern synthesis' of natural selection and population genetics. In this he was building on the work of Fisher, among others, who was a good friend. Though they were at opposite ends of the political spectrum – Fisher a conservative, Huxley an admirer of central planning and the Soviet Union – they shared an enthusiasm for eugenics. 'I was sorry not to see you at the Sterilisation meeting yesterday,' Huxley wrote to Fisher in September 1930, for example. 'We must not let Blacker lose his nerve about the sterilisation proposals,' Fisher replied a few days later, a reference to a paper by Huxley and psychiatrist C.P. Blacker recommending the legalisation of voluntary sterilisation so that 'especially poor people' could get themselves sterilised as part of a plan to reduce the number of the mentally deficient being born. Though ostensibly voluntary, this sterilisation was going to be cajoled if not coerced. As Huxley said in his Galton Lecture in 1936, 'The lowest strata, allegedly less well-endowed genetically, are reproducing relatively too fast. Therefore birth-control methods must be taught them; they must not have too easy access to relief or hospital treatment lest the removal of the last check on natural selection should make it too easy for children to be produced or to survive; long unemployment should be a ground for sterilization, or at least relief should be contingent upon no further children being brought into the world; and so on.' Even in 1930 certain unhelpful people were warning about this being a slippery slope to cruelty if not worse. But Britain was lagging behind both the United States and Germany in its enthusiasm for such eugenic policies – so Blacker must not lose his nerve.

This eugenic background is relevant because it surely coloured the perspective of both men when they watched birds being careful and discerning choosers of mates. Yet, as I say, Fisher's sexy-son insight was if anything anti-eugenic. He and Huxley corresponded throughout the 1930s on eugenics, but as far as I can find never about sexual selection. Which is strange because Huxley was instrumental in burying Fisher's vital contribution to that debate, using faint praise and neglect.

In 1938 Huxley wrote a long essay for the *American Naturalist* on Darwin's theory of sexual selection, followed by another as a chapter in a book. He now dismissed all female choice theory in a few dogmatic sentences: 'Many conspicuous characters (bright colors, songs, special structures or modes of behavior), to which Darwin assigned display function, have now been shown to have other functions,' he asserted. Again, this would have been news to those who had watched birds closely, such as Selous. As for his friend Fisher, Huxley mentioned the runaway process in a single sentence and then changed the subject. And here came the old Wallacean trick of subsuming sexual selection under natural selection: ornaments 'need no special category of "sexual selection" to explain their origin,' said Huxley. Darwin, he declared, 'persistently attached too much weight to the view that bright colours and other conspicuous characters must have a sexual function'. And finally: 'Darwin's original contention will not hold.'

The psycho-physiological effect of sexual excitement

Huxley grasped at any explanation for male display as long as it was not female choice. The majority of Darwin's examples were simply of ornaments that are useful for males in threatening other males, he said. Exactly why a Peacock should be intimidated by a decorated train covered in eyes, rather than say brute force and long spurs, is left unclear – as is why Peacocks fold away their trains when they fight with other males. In this Huxley echoed Richard Hingston, a doctor based in India who had been the medic on George Mallory's final and fatal Everest expedition in 1924. Hingston published a book on 'the meaning of animal colour and adornment' in 1933 in which he called bright colours 'intimidating machinery'.

Huxley argued that other male ornaments served to remind birds to mate with members of the same species, even though dull-coloured species seemed to manage this quite happily. He also conjectured that sometimes the purpose of bright plumage in males was to deflect predator attacks from more valuable females, invoking the German biologist Oskar Heinroth's description of the spectacular display of a male Bulwer's Pheasant at London Zoo – a display that in the wild renders the male highly conspicuous to passing leopards. So a male is a mere decoy, wearing a sign reading 'Eat me!' Really? A moment's thought would remind any scientist that such self-sacrifice would struggle to evolve: if predators keep killing the most colourful males, how do their genes get passed on? A similar argument had been made in the 1880s by the Polish ornithologist Jean Stolzmann, who thought that in most birds more males were born than females and the bright colours of males served to ensure that they were culled – which would quickly weed out the coloured males, surely? In any case, the premise was false: sex ratio at birth in most species is close to 50:50. (The Ruff is a surprising and intriguing exception but in the other direction: its populations seem to be female-biased, with about 70 per cent of birds being born female.)

Then in the next breath Huxley argued that the functions of ornaments also include warning the predator that an animal is toxic to eat. So the bright feathers are sometimes signs reading 'Eat me' and sometimes signs reading 'Don't eat me'? Make up your mind! It is of course true that some brightly coloured caterpillars, snakes and fish use bright yellow and black colours to warn predators not to eat them, as Darwin was well aware, and the same may be true of some birds, but the vast majority of sexually ornamented species including Peafowl are eaten with relish by predators.

The infuriating thing about Huxley is that he comes close to getting it right, and in doing so repeatedly contradicts himself. He writes:

> In other cases, however, the advantage probably lies in the psycho-physiological effect of sexual excitement. I would, for instance, suggest that this is the explanation of the constant sparring of ruff and blackcock on their mating-grounds, which all observers agree very rarely develops into actual fighting.

So he knows that a lek is a place of love not war, and he knows that the females attend the lek to mate with males, yet he seems to think that a male displaying at females on a lek has little or no implication for mate choice. His excuse was that in his grebes and divers male display served to induce in females 'a psycho-physiological state of readiness to mate, irrespective of any possibility of choice', because it happened only once birds had paired up, not when they were selecting mates. His grebes were apparently strictly monogamous (ah, but were they? Watch this space), yet they indulged in dances involving the ecstatic passing of weed while treading water in elegant if unnatural postures. Since they had already paired up before doing this dance, he could not see how display played a role in mate selection. Its purpose was to synchronise the birds' physiology so they were both ready to sit on eggs, he concluded.

This notion dated back to Darwin's old enemy Mivart, who said 'the display of the male may be useful in supplying the necessary degree of stimulation' to the female's nervous system to make her willing to mate. Huxley adopted the term 'epigamic' display for this idea, a word coined by a champion of Darwinism named Edward Poulton in 1890, in a book about the colours of insects. Poulton, seen as Darwin's most prominent disciple in the early twentieth century, had started out as a fan of female choice, but he later moved away from this position, saying that sexual selection is 'entirely subordinate to natural selection'. Then Vernon Kellogg, a Stanford University entomologist who taught President Herbert Hoover, made the same argument in a book in 1907. Colours and songs 'probably do exercise an exciting effect on the females and are probably actually displayed for this purpose,' he conceded. 'But does this in any way prove, or even give basis for a reasonable presumption for belief in a discriminating and definitive choice among the males on the part of the female?'

Well, yes it does. Karl Groos had made a similar point, but gave the game away: 'A kind of unconscious choosing does take place which is in a peculiar sense sexual selection, for the female is undoubtedly more easily won by the male that most strongly excites her sexual instinct.' Yes! How could Mivart, Kellogg, Poulton and Huxley not see that stimulating a female into a state of readiness to mate and being chosen by a female as a mate might effectively be the same thing? Remember in Black Grouse and many other lekking species a female deliberately flies towards a gathering of displaying males a day or two before she begins laying eggs. She takes the initiative in exposing herself to the possibility of being impressed by a display. In Argus Pheasants, a female deliberately visits a male's dance ground. She is plainly already fairly strongly motivated to mate before she has seen any showing off. She comes to the lek or the dance ground and plays hard to get until she sees a particularly good performance: that's choice, and it is up to you whether you see it as a process decided by the male's display or the female's discrimination. From the vantage point of the twenty-first century, the theory of epigamic selection looks like a sexist attempt by crusty old misogynist men (Selous excepted) not to confront, let alone admit, the possibility of female lust – except when it is aroused by men.

I have been a bit harsh on Huxley. By the time he came to write *Evolution: The Modern Synthesis* in 1942, he was prepared once again grudgingly to admit that in a few cases – the Ruff and the Black Grouse are once more the two examples he cites – male ornaments and bright colours are used by males in both threat displays aimed at other males and seduction displays aimed at females. He even added that in the case of the Peacock's train, the 'display function appears to be the only one'. But he devoted a single page to the topic of sexual selection in a 645-page book on evolution, which seems rather mean. As Helena Cronin has pointed out, the other scientists who revived Darwinism in the 1930s and 1940s barely mentioned the topic at all. In the influential books by George Simpson and Theodosius Dobzhansky, the phrase does not even feature in the index. Despite Selous confirming Darwin's conjecture that female choice does happen and Fisher giving a logical reason why, the idea of sexual selection as an explanation for beauty in birds was once again dead and buried by the mid twentieth century. Selous is not in the index or the bibliography of his friend Huxley's huge book.

As the centenary of Darwinism dawned in 1959, natural selection was still in vogue, but sexual selection was the embarrassing uncle at the party. Yet in human society, sexual liberation was stirring, and the idea of women as passive recipients of men's decisions was fading.

'Please, God, bring me a mating today!'

'A lek is like nothing else in nature.'

'The dark blue, tremulous neck expanded to more than twice the normal girth.'

'A clockwork cackle announces the arrival nearby of a cock Red Grouse.'

'Up close the resemblance to a pair of sea anemones is striking.'

'The first female takes offence at this and flares her tail.'

'Wonky Tail gets his name from the way his tail feathers are always at an angle.'

'The birds leap at each other, try to grab each other's feathers.'

'Yearling males have speckled brown feathers on the head, neck and back.'

'Battered and bruised at the end of the season.'

'The female Grey (or Red) Phalarope is brightly coloured.'

'The duller-coloured male sits on the eggs.'

'Temminck's Tragopan shakes loose a lappet of electric blue skin beneath his throat.' (Photo by Jonathan Pointer)

'The iridescent blue Monal of the high Himalayas.' (Photo by Carolien Hoek)

'The song of the Curlew is a thrill to the human soul.'

'The male Lady Amherst Pheasant opens a ruff of concentric patterns around his real eye.' (Photo by Henry Koh)

'In Puffins both sexes have colourful beaks in the breeding season.'

5.

Long Odds

But neither breath of morn when she ascends
With charm of earliest birds, nor rising sun
On this delightful land, nor herb, fruit, flower,
Glistring with dew, nor fragrance after showers,
Nor grateful evening mild, nor silent night
With this her solemn bird, nor walk by moon,
Or glittering starlight without thee is sweet.

JOHN MILTON, *Paradise Lost*

6.30 a.m., April, the Pennine hills

The lek is in full swing now. Messy Shoulder, the bird whose patch is closest to the hide (and who has some out of place feathers on his right wing shoulder), is picking a fight with old Wonky Tail. Just next to him is a male missing a feather from his fan of white bum feathers. He too is a fairly central, successful male, but he is not the top cock. On the other side is Black Bar, with the distinctive narrow black line across the top of one of his white bum feathers.

When fighting breaks out the birds make what Selous rendered as the 'Choc-ke-ra-da' call, a sort of irritated bleating. Their neck swelling shrinks. Each male rushes over to square up to his neighbour whenever the latter approaches the invisible boundary of his patch. This first level of fight (threat) is common and often matters will end there. But sometimes the duel escalates to the next level in which the males spring back and forth in what looks almost like a choreographed way to avoid each other's stabbing beaks. Most of the time, after a few parries, the squaring up ends with a cautious disengagement as the birds slowly turn away from each other. Only rarely does the fight escalate to the third level where the birds leap at each other, try to grab each other's feathers and wrestle each other ferociously with flapping wings and slashing claws, feathers flying.

On this occasion I watch Wonky Tail disengage from his confrontation with Messy Shoulder apparently because he wants to pick a fight with Touching Combs instead: he goes straight from one contest to the other. Messy Shoulder is not bothered, almost as if he realises it's

Two males squaring up for a fight.

Touching Combs's turn to fight Wonky Tail. Indeed, now I come to think of it every confrontation is dyadic; there never seems to be a three-way fight. Sometimes a third male will stand watching a fight between two of his neighbours, waiting his turn to have a duel with one of them.

Such fights are rare, but that Black Grouse are in deadly earnest when they do fight is shown by the wounds that are sometimes inflicted. And by the appearance in Scandinavia of occasional 'Rackelhahns'. These are hybrid males between Black Grouse and larger Capercaillies. Bigger, blacker and with a longer reach than Black Grouse, but drawn to their leks, these hybrid birds can be lethal on a Black Grouse lek. It's horrifying to watch footage in which a Rackelhahn starts a face-off with a male Black Grouse but then uses his longer reach to grab the wretched Black Grouse by the head and peck him to death.

Wonky Tail squaring up to Touching Combs.

In one film taken in Norway a few years ago, the Rackelhahn continues beating the dying Black Grouse till he eventually decapitates it and carries off the head, then grotesquely returns and copulates with the headless corpse. Fortunately, the Rackelhahns are probably sterile like many hybrid animals. Scandinavian hunters sometimes kill them to prevent the slaughter.

Today, with no such murderer on the loose, the central male, Black Spot, stands aloof, watching the two combatants, Touching Combs and Wonky Tail. To the naked eye there is nothing that distinguishes him from another bird apart from that tiny black spot. Perhaps his wattles are a little brighter, his breast a little bluer, his tail a touch longer, but it's hard to discern.

As I mentioned in chapter 1 when regretting the persistent Arthurian myth about the lek, fights between neighbouring males do not end with victors and vanquished, at least not in the sense of one male flee-ing for his life. The two birds square up, box and parry a bit, rarely hit each other with wings and feet, and then return to their own territories on equal terms, each king of his own castle. A dominance hierarchy it is not. The central males, with their fixed positions, face their approxi-mate equals and turn home advantage into a strong factor. That's the peculiar thing about territory in the animal kingdom: it bestows limited, arbitrary dominance on the owner. Birds and butterflies, fish and squid all show this curious phenomenon of home advantage. My old friend Nick Davies once watched Speckled Wood butterflies defending patches of sunlight in a wood and saw them respect each other's rights: possession was nine-tenths of the law. By quietly setting up a male in a patch of light owned by another, he tricked two butter-flies into thinking they both owned the same patch of sunlight, resulting in a far more furious and long-lasting fight.

In Finland, after filming 1,082 fights between Black Grouse males on leks, Anni Hämäläinen and her colleagues came to believe that they could tell which bird was the winner at the end of the fight. Surprisingly,

Black Spot was the male who got most of the matings in 2023.

it was the one that turned away first, whereas normally in animals turning tail is an admission of defeat. 'We believe that Black Grouse fights do not terminate with the subordinate turning away from the dominant. Instead, dominant males turn first and present their tail towards the subordinate.' This is Hämäläinen's interpretation and I am not entirely convinced, but her argument is that presenting your vulnerable back to your opponent, while spreading your lyre tail and resuming roo-kooing, is a sign of your confidence that he will not risk attacking you. And it is certainly true that the other bird usually does not seem to take advantage of his rival's turn, but rather looks a little relieved that the contest is over. In Hämäläinen's data, the 'winners' of these fights, under this definition, were actually slightly less bright blue on the neck but apart from that there was no feature of the plumage or body that reliably predicted who would win a fight. Central males on the lek were not especially likely to win fights by this definition either, though they do have to fight more often to retain their top territory. Rather this confirms the equal terms on which the central males live with their neighbours. As I shall show shortly, it is far from clear that the ability to win jousts is a factor that females take into account: fighting tends to die down when females appear.

However, back in 1991 Rauno Alatalo and his colleagues Jacob Höglund and Arne Lundberg had found that the males that were dominant in winter flocks were also the most successful in mating. In the dummy-female experiments, males that won fights over female dummies at territory boundaries were also more successful. They pointed out that males 'tear feathers from each others' tail ornaments in combats, and attractive males always had undamaged tails'. Moreover, the dominant and successful males tended to live longer. So it does seem to be true that females are generally mating with the dominant males, rather than just the fanciest.

'The year of peak lekking effort tended to occur in the last year the male was alive'

If you are a male Black Grouse the statistics are brutal. According to the Finnish study, more than half of all males will never copulate at all during their lifetimes. Of the ones that do mate, only a few will get to do so more than a few times, and of these lucky winners most have just one successful year. That's because three-quarters of those that hold a central territory and mate many times in a season are dead within the year. The ordeal of holding off rivals, pouring your energy into incessant display and being set upon every time you mate – and being unable to feed for several hours in both morning and evening – is just too much for your body to bear.

The data collected by Matti Kervinen, Christophe Lebigre and Carl Soulsbury from Finnish Black Grouse leks over ten years seem to suggest that it's the behaviour of the males at the lek, more than their mere appearance, that crucially determines their mating success. In their study, just 12 of 164 males accounted for half of all matings and 85 males never mated at all. One male mated 21 times in one year and 11 in the next, making 32 clutches he probably sired in his life, or up to 250 eggs in total.

What determines success in any one year? Sure, the most successful males tend to be the ones with longer lyres in their tails and greater body weight but that is partly because these are the older males. The yearlings and two-year-olds are usually significantly lighter and shorter in the tail than those that are three or older. On average the birds weigh 10 per cent more going into their second spring than their first and 5 per cent more again going into their third – that is, as they approach their third birthday. After that their weight stays roughly the same. Tail length increases from about 190 millimetres in yearlings to 220 millimetres in two-year-olds and then is much the same in future years with a hint of shortening again after five years if they survive that long. Eye comb size goes on increasing each year till the birds are four years old, whereas the measured reflectance of the blue chroma on the neck seems to peak at two years old and decline a little after that.

Measures of behaviour are also influenced by age. In the Finnish study, the yearling males on average spent only about 20 per cent as much time at the lek as the male that spent the most time there, and two-year-olds about 70 per cent. Up to the age of five, males grew steadily more pugnacious each year, so that by that age they were spending about a third of their time at the lek fighting or squaring off with rivals. And as each year went by a male would, on average, get closer to the centre – or the lek centre would get closer to him – at least until he was four years old.

Thus, controlling for age reveals a subtle effect. Once a male reaches the age of three his chances of having a successful mating year – and remember he usually only gets one spring at the top – do depend to some extent on his weight and his tail length in that year. But the sooner he can establish himself on a lek, the sooner he gets to the centre and stands a chance of winning the genetic jackpot one year. Progress to the centre depends on whether he can establish a position, how frequently he attends, how often he fights and – perhaps – how vigorously he displays. These behavioural factors seem to matter more than how long his tail is or how much he weighs in deciding whether he gets to be chosen by females. That's probably because behaviour reflects the current state of the bird whereas plumage is determined by his state when he was growing the feathers a few months before. As Kervinen, Lebigre and Soulsbury write: 'Sexual selection in Black Grouse operates primarily on male behaviour.' They add that 'morphological traits may act as additional cues to supplement female choice'.

So the fancy plumage is a necessary but not sufficient feature of a successful male: behaviour matters more. The Finnish team also found little sign of males declining in their ability to win a mating spree in old age – at least on average. Indeed a footnote to a chart in one of the team's papers tells one eventually triumphant tale for an especially elderly male: 'The high mean annual mating success … at age 5+ of the males that began lekking at age 1 is largely affected by one male that had 15 copulations at age 6.'

Yet when they looked more closely, the Finnish scientists found that the average concealed a secret. Although the average bird was as successful at five as he was at three, for individual birds, there was indeed a striking senescent effect. On all the seven traits that the Finnish team measured – weight, tail length, blue chroma, eye comb size, lek attendance, time spent fighting and distance from the lek centre – there was a decline from peak performance after the individual's best year, whenever that was. It was just that the peak year came at different ages for different individual birds.

It seems that a male Black Grouse will put in one year of enormous effort when he is ready, and that he exhausts and debilitates himself so much that he cannot match it again after that. According to the study, 'The year of peak lekking effort tended to occur in the last year the male was alive, especially in short-lived males.' This is the third year I have watched Wonky Tail, who must be at least five years old given that he had a fairly central territory in the first year I watched him, and it is the first year in which he has achieved a (solitary) mating; I think I detect a heightened pugnacity in him this year too. If this is his year of peak effort, I fear he may not be back next year. It looks from the Finnish data to be the case that if the birds effectively get one shot at mating, then perhaps the year of peak performance kills them whether it results in a mating spree or not.

This fascinating pattern, an insight that would not have been suspected without diligent and dedicated observation over thousands of hours and ten years, was set out in a paper the Finnish team

Feathers are flying.

published in 2015 in the *American Naturalist* journal. It was to be Rauno Alatalo's last publication, as he died of cancer during its preparation, and the paper is dedicated to his memory. At the extreme ends of the spectrum, there seem to be two strategies that Black Grouse can instinctively adopt: get off to a fast start, peak early and die young; or start slow, peak later and live longer. It is as if the birds unconsciously sense when they are in with a chance of mating and put everything into that year's effort.

'Mere uncouth violence she does not appreciate'

My musing on this life-history gamble is interrupted by a flurry of activity at the lek. A couple of yearling cocks, distinguishable by their much browner backs and shorter tail lyres, land at the edge, causing a brief flurry of flutter-jumping among the males who seem to mistake them briefly for incoming females. The yearlings move rapidly in towards the centre, drawn it seems by the magnetism of the densest part of the lek. But each male whose territory they cross runs furiously at them and the two youngsters find themselves assailed on all sides and speeding right through the lek to the far end, taking off to escape the lunges of the residents. Here is no stately squaring off between 'dear enemies'. The intruders are hit hard if they slow down.

Yearlings do very occasionally get to mate: in Finland 13 out of 193 yearlings did manage to mate in their first season, mostly in three mating seasons when the population was expanding again after a period of decline. At such times new leks can form consisting only of yearling males and some of these will attract females, giving the larger and more advanced yearlings an early chance to mate. A mile to the west of where I am sitting, there is a new lek this year in a sheep pasture with eight males, some or perhaps all of whom are yearlings. These new leks are not very stable, the males sometimes abandoning them for larger leks later in the morning, and their locations tend to move in subsequent years. There is a dilemma for a yearling cock: whether to be a big fish in a small pool or vice versa.

As the season progresses, the yearlings that come to the main leks will get bolder, capitalising on the exhaustion of the primary males. When I visit the lek in mid May – getting up at 3.30 a.m. to beat the sunrise – I will find a different scene from late April. The central males may be missing chunks of feathers on their napes or backs and may have scars on their faces. Evidence from the Finnish study shows they will probably have lost weight and their parasite load will be higher, especially in the most central males. Displaying will certainly be more sporadic, and periods of quiet when most of the males are resting with their combs shrunk and their heads sunk into their shoulders are more frequent. In May the arrival of yearling males is still greeted with apparent fury by the residents, but the yearlings seem to know that they can now challenge their elders. At this later date the youngsters occasionally will turn on their attackers and flare their tails at the edge of the lek, trying to stand their ground. A vicious fight may ensue before the younger cock realises he is still not quite ready for prime time, and turns tail. Sometimes the young male does manage to find a spot far from the centre where he can display unchallenged for a while. Another paper by Carl Soulsbury and his colleagues documents that my anecdotal observations here are not misleading. They measured an increasing attendance rate of yearlings at leks as the season progresses and an increasing frequency with which they are prepared to fight. At the leks they studied both attendance rate and fighting rate for yearlings roughly doubled between the start of the season and the latter part.

Mortality among males is highest immediately after the lekking season, when hunger, sickness and predators take a high toll of males. Battered and bruised at the end of the season, exhausted and sickened, the central cocks nevertheless find that the game of thrones can be worth it, genetically speaking. The central cock on a big lek like this may achieve at least as many matings as he has rivals if he gets to be the only male that mates: that means maybe fourteen or even more on this lek. This is because the bigger the lek, the more attractive it is to females and the more likely they are to mate while there. So a lek of fourteen may attract as many or more females and the top cock might mate with almost all of them. One morning I saw seventeen females on the lek at once, outnumbering the males: a chaotic and confusing morning it was too, with males uncertain which way to turn when displaying. No matings happened that day, however.

That large leks are more attractive to females has now been proved: for every extra male at a lek there are approximately 1.75 more copulations in a season in Finland. The female preference for large leks possibly explains how leks came to be: by their preference for large, dense leks, females are quite literally breeding the lekking habit into their mates as surely as dog breeders breed retrieving habits into Labradors. Joining a large lek such as this is a pretty futile quest for a young male. He has a small chance of making it to top cock even if he lives to five or six. But if he does, he wins the jackpot.

In any case there is no alternative. In a lekking species, males are bred to lek – by females. The idea that leks exist because females created them took a long time to emerge within science. Over the years scientists came up with lots of explanations of lekking behaviour. Perhaps the males are trying to reduce predation by gathering in flocks while they display so that they can all share in vigilance. Perhaps they are hoping to amplify their calls so that they reach females further away. Maybe they are forced to gather by the nature of the habitat they live in, there being only a few spots that are available for display. These are all male-centred arguments and in a ground-breaking paper in 1983 Jack Bradbury of the University of California, La Jolla, who had studied Sage Grouse, found serious flaws in all of them. Instead, he suggested either that males gravitate to spots where females are likely to show up – the so-called hotspot model – or that females prefer large gatherings of males because it enables them to compare males and choose the fittest.

Looking out at the lek, I am all but certain that the hotspot model won't work. The site of the lek is always unremarkable and this one is no exception. It's on a very gentle ridge of ground, but not at the highest point. The ground is flattish but not completely level. It's in sparse, low vegetation but still there are clumps of rushes. Every Black Grouse lek I know is like this and usually well away from the main feeding areas of the females. At this time of year the females are feeding in meadows and bogs, maybe also in woods. They don't just happen to be at this lek site; they come to it specially – by air – to mate. Sage Grouse do feed at the lek, which takes place in sage brush, the species' main food plant. But Black Grouse find very little to eat at the lek. If they are lekking on a frozen lake in Finland, there's no food at all.

However, in 1988 Bruce Beehler and Mercedes Foster of the Smithsonian Institution, who had studied birds of paradise and manakins respectively, argued for a different explanation of the evolution of lekking behaviour: the 'hotshot' model, in which males cluster around the sexiest male, because that's their best – though still small – chance of getting a mating.

Yet Beehler and Foster were not persuaded that females really have much freedom of choice at leks. There's too much evidence in many species of males interrupting other males that try to copulate, in effect controlling who gets to mate, and too little evidence of any useful differences in plumage or behaviour between males for the females to use as cues to the best male, or so they argued. Thus males are mostly deciding among themselves who is allowed to mate, they suggested, and female freedom to choose has been overstated. So the old Victorian naturalists were right? A lek is a jousting tournament and what interests the females is the outcomes of male fights to decide who is dominant. They are passive beneficiaries of a male battle. 'Field evidence indicates that females are more conservative than they are choosy,' the two scientists wrote: females choose the same mate year after year and young females tend to copy the choice of older females. Because of copying, an 'initially successful male acquires more and more mates with each passing year, eventually achieving hotshot status'.

I spy a snag. Unlike birds of paradise and manakins, Black Grouse rarely survive long enough to get two shots at being a hotshot. Instead hotshot status is instant: achieved over (and for) just a few days, the mating week or so within a single season. Remember the Finnish statistics: three-quarters of males that mate are dead by the next season. Nor does dominance feel like the right way to describe a lek where each male is deadly dominant in his one small patch but pathetically subordinate the moment he crosses the border.

Moreover, if Beehler and Foster are right and it's the males who decide among themselves who will mate with each female, then you would expect males to fight more as females approach. The opposite is true. Far from intensifying, the fighting dies away somewhat when a female turns up. Anecdotally I have often noticed this, and so did Selous, who wrote, 'The presence of the hen on this occasion has not brought about a combat but rather diverted it', and noted that for display 'alone as a spectacle the female bird has eyes; mere uncouth violence, though effective enough for us, she does not appreciate', though how he could tell what was going on inside her head is not made clear.

Scientists have now actually measured this effect. In a study of a Sage Grouse lek in Wyoming, Sam Snow and colleagues used a sophisticated statistical technique called a relational event model to analyse 1,506 interactions among males leading up to 72 'copulation events' and work out exactly what happened moment by moment. Their findings vindicate Selous: 'Our analysis reveals that fighting's primary function is not to impress females. Indeed, males are less likely to start and more likely to leave fights with females present.' A male's mating success depends in part on his tactical ability to avoid allowing fights to interrupt his courtship of a female. Nor did aggression explain the failure of many males on the lek to mate, as Beehler and Foster argued: plenty of males remain on the lek undefeated but unmated. Snow's conclusion is emphatic: 'We recovered no evidence that fighting is directly attractive, informative, or functions in an additive way with display to contribute to the solicitation of copulation by females. In fact, males are less likely to start fights and more likely to end them sooner in the presence of females on the lek.'

'The male gives a buzzy, snarling two-syllable call in flight'

In any case, I think Beehler and Foster had overlooked something simple: the Fisher runaway effect. Neither Fisher nor those who tested his idea in the 1980s (see chapter 4) appear in the list of references at the end of the Beehler-Foster paper. The runaway effect is a version of the hotshot model and it explains copying very easily: in a polygamous species each female is (unconsciously) desperate to ensure that her sons are attractive so they stand a chance of winning the genetic prize. She must therefore follow the local fashion in mate choice, and this will have been bred into her and continues even when the difference between the plumage of the males on the lek is trivial. Copying surely explains why males are forced (by females) to aggregate. Once copying becomes established, females are going to exhibit a degree of gregariousness in the mating season and are going to be drawn to the largest aggregations of males so they have the best chance of making the same choice as their rivals.

I come back to a very simple point. Males gather at leks because females choose to mate at leks. Hotshot males become hotshot males because females have an inherited habit of choosing to mate with the same male as each other, right in the centre of the lek. What's more, as Snow's data suggests, females have been selecting males that do not just fight each other but also tolerate each other's close proximity. If anything they have bred reduced aggressiveness into the species. I say again, a Black Grouse lek looks awfully like a team: a team of rivals, sure, but still a team. For all the posturing and occasional fighting on leks, there's not many male birds that would tolerate another male within a few metres during the breeding season, let alone lots of them. When lekking ends at breakfast time the birds on my lek will be found feeding together in a field, as comrades. In a lek further up the valley that has eight males on it this year, sure enough when I pass by a meadow near there on my way home later, all eight are feeding together as a flock, the best of friends. An hour before they were ready to maim each other.

Later in the spring, I observe the same phenomenon in Arctic Norway. The species I am watching there is the Ruff, the lekking shore-bird in which the males grow colourful feather stoles around their necks each spring. For much of the daytime the males are away from the lek, feeding in fields and bogs. But what strikes me is how in all three of the different places I come across Ruffs one day – on a sheep pasture, in a sedge marsh and in a hayfield – they are in a tight flock, moving together. This is unlike every other shorebird in the area. In each case the flock consists of members of a single lek. In what passes for dawn at these latitudes, they will be on the lek, knocking seven bells out of each other for the benefit of females, known as Reeves. The old Latin name of the species, *Philomachus pugnax*, means 'belligerent lover of fighting'. But here, a few hours later, they are best friends. As I say, a lek is really a team effort.

In the coastal forests of Ecuador lives a bird that is so bizarre in appearance no *Monty Python* sketchwriter would think of inventing it. The male Long-wattled Umbrellabird is a black, fruit-eating cotinga with a strange, smooth, black dome of feathers that looks like a helmet over its head and beak. From its breast dangles a floppy, sausage-shaped appendage, covered in scaly feathers that extends to almost twice the length of the body and appears both inconvenient and silly. It would appear that females have played a cruel joke on males in this species, making them look ridiculous and comic. Long-wattled Umbrellabirds lek, albeit in a rather dispersed way. One team of scientists led by Luke Anderson of Tulane University studied them and recorded that – as in Black Grouse and Ruffs – they often leave the lek together to forage as a group: 'During periods of high lekking activity, male umbrellabirds depart the lek in highly coordinated groups and maintain larger off-lek social groups relative to periods of low lekking activity.' This is just like Black Grouse and Ruffs. Through careful observation the researchers ruled out various explanations for this pattern, such as that their food was concentrated at that time of year, that they just happened to need water at the same time, that it helped them avoid predators and so forth. Instead, the way the males but not females leave the lek together, stick together while off the lek and return to it together suggests that the point of this behaviour is to coordinate the team game of being at the lek simultaneously and displaying as a group, the better to attract females.

In some species of bird the teaming up of males at leks has gone still further. In certain manakins in South America, two or more males display cooperatively, doing coordinated dances for the visiting females. Richard Prum has described how pairs of male Golden-winged Manakins dance in a collaborative way: the first male waits on a log for the second to arrive then leaps up into the air to be replaced by the second male, before the roles are reversed. In Blue Manakins, the cooperative dancing has become even more elaborate. Four or five males perform a carefully staged series of 'backward-leapfrog' displays: 'After the female has landed on the display perch occupied by the males, the male who is closest to her leaps upward and hovers in the air in front of her with his red crown fluffed. While hovering, the male gives a buzzy, snarling two-syllable call in flight and then flutters back down to the perch, taking a position farther away from the female. Meanwhile, the second male slides forward along the perch toward the female, then leaps up and performs the same display as the first.' This may be repeated up to a hundred times before the female is induced by one of the males to mate.

Reading these accounts of manakins, I rethink what I see at the Black Grouse lek, trying to imagine that these squabbling males are mere allies of the top male, assisting him with their displays in seducing a visiting female then heading off to feed alongside him later, muttering 'Glad to have helped you, old chap.' What's in it for them? Are they hoping to inherit the top spot? Probably. But studies of manakins show that it can take five or more years for a beta male to inherit the alpha's role, and in one cohort of twenty-one Lance-tailed Manakins born in the same year, only five made it to alpha status and achieved any matings at all over nine years. Or are the males, as Prum suggests, mere victims of generations of female selection for joint dancing? 'Obligatory male-male cooperation is a total loss for the vast majority of males. The only reason it can happen is that females are completely in charge. Males have no options because there is no other game in town.' Females, in other words, can make males into team players if they want.

In turkeys, too, a group of brothers will usually team up to see off other groups of males and display together to a female. Given another fifty thousand years, maybe Black Grouse leks too will become dominated by similar team games.

Could these teams be relatives? If they are brothers, half-brothers and cousins, sired by the same few highly successful males, their collaborative spirit might be a case of 'kin selection', the evolutionary phenomenon that explains ant colonies and beehives: namely, getting your genes into the next generation via your siblings might be a better strategy than trying to breed yourself. In Peacocks, half-brothers do tend to join the same lek more often than unrelated individuals even if reared in captivity. Yet in Black Grouse in practice there is sufficient dispersal between leks in the first year that the birds on a lek are not all especially closely related. The Finnish study found that most male Black Grouse had no closely related neighbour on a lek, and nor did related neighbours go easy on each other when fighting. So there seems to be no attempt to team up with brothers against others.

'Females who would wander into the lek to choose their mate'

Even weapons used in fighting might actually be subject to mate choice as well. An example might be the outsize antlers of *Megaloceros giganteus*, the European Giant Elk of the ice age, which is also sometimes known as the Irish Elk because of the tendency of its half-fossilised remains to turn up in Irish peat bogs. It was once thought that the Giant Elk's massive and unwieldy antlers, up to four metres (twelve feet) across and forty kilograms (eighty-eight pounds) in weight, must have been the cause of its extinction. Chased through a forest by cavemen with spears, the animals supposedly could not fit between the trees and succumbed to their pursuers, victims of too much sexual selection for fighting ornaments. This notion no longer commands much support among scientists because in some places the animal died out as the ice age ended, before human beings got there. It died out in the British Isles about nine thousand years ago, surviving a little longer in parts of Siberia. But in any case they lived in open semi-wooded steppes with generous gaps between trees, and the death rate in winter in at least one Irish bog seems to have been highest in weaker males with modest antlers.

But were the antlers used for fighting or display? It is not even clear that they were outsized. Given that antler size is disproportionately greater in larger deer anyway, in non-linear fashion, the Giant Elk's antlers are just about the 'right' size compared with say Fallow Deer, their closest relatives. It is more the case that Moose, which are about the same size as Giant Elk, have antlers that are undersized on this measure. Yet a study of the stresses that their antlers and skeletons could take concluded that the Giant Elk's antlers were only just capable of withstanding fights and were better able to resist twisting than pushing movements.

I don't think it can be ruled out that the great size of the Giant Elk's antlers might have been the result of female choice rather than male fighting or even perhaps that the stags lekked, gathering at special places in the rut to strut before discerning hinds. Professor Adrian Lister of London's Natural History Museum thinks so: 'In this display of strength, bigger antlers were probably more intimidating to other males and more desirable to the females, who would wander into the lek to choose their mate.' After all, Fallow Deer do lek, the males gathering in groups and calling to the females from small territories rather than herding harems like Red Deer. And Fallow Deer do have rather fancy, broadly palmate antlers, not unlike smaller versions of Giant Elk. As soon as we can bring the Giant Elk back from extinction, using gene editing, we can set them free and find out – maybe.

Here also lies a potential explanation of why the size difference between the sexes is much greater in birds and mammals where males fight a lot but smaller in those that display to females a lot. Black Grouse males are bigger than females but not by a large margin – males weigh about 20–30 per cent more than females. Not like Elephant Seals whose adult males can weigh three times as much as females and among whom fighting over females is the rule rather than trying to persuade them to mate.

Sometimes the need to display energetically seems to have led to males becoming smaller than females, not larger. More than forty years ago I spent a summer in western India studying a rare bustard, the

Lesser Florican. For some reason a grant from the American taxpayer needed spending that year or it would be lost and nobody else was available, so I got the gig along with two cheerful and brilliant Australians, John Woinarski and Rob Magrath. A kindly, pony-tailed, gun-toting maharajah who was passionate about conservation lent us a dilapidated palace to live in for free and provided us with a driver to explore the meadows and fields of the Saurashtra peninsula in an attempt to census the vanishingly rare birds. About the size of a chicken, the florican has long legs and a long neck, like most other bustards. The females are brown and well camouflaged in the grass-lands where they live. The males have yellow legs, a black neck and belly, fringed with white along the wings. From its cheeks extend back-

Male Lesser Florican in western India.

wards some fine, black, curved plumes that are wider at the end like long, thin commas – plumes that were popular additions to Edwardian hats. It is not a lekking species, since the males are widely dispersed, but they take no part in hatching the eggs or raising the chicks, so all their effort goes into trying to attract females.

In the monsoon season each male selects a few spots in a grassland or crop and moves between them, displaying several times at each location, by leaping about six feet into the air, throwing back his head, flapping his wings with a loud rattling sound and then dropping back to earth. Over the course of three months, in the hot, humid monsoon, we found sixty-nine displaying males and soon realised that they preferred to display in cloudy conditions, either to avoid the heat of

In the breeding season, Lesser Florican males leap in the air several hundred times a day.

the sun or to avoid becoming a target for eagles. By watching one male at a spot called Harshadpur for more than forty hours over eight days, my two colleagues and I were able to calculate just how often the bird did this 'display leap'. When displaying it jumped once every sixty-one seconds on average and spent 38 per cent of the time displaying. That meant it would leap around four hundred times a day. The purpose of the display is to attract females arriving on the breeding grounds from elsewhere in India. Here's the curious thing: the Lesser Florican is the smallest species of bustard, and the only species of bustard in which males are significantly smaller than females. On average a male's wing is 188 millimetres long, a female's 232 millimetres. This small size would not help males in fights with each other, which are rare, but would be an advantage when it comes to hoisting yourself six feet into the air every sixty seconds in hot weather for several hours. The next smallest bustard, Savile's Bustard, which lives in the Sahel region of Africa, does not have an aerial display and males are slightly larger than females.

Today is a good day for Black Spot. He mates with six females, one after another. For one of them, which I can recognise because she is missing her left eye, this is her third visit to the lek in three days, but it's the first time she has mated. That seems to be typical. The sperm is stored for long enough to fertilise each egg just before it is laid.

6.

Tales of Long Tails

Since Nature hath inviolably decreed
What each can do, what each can never do;
Since naught is changed, but all things so abide
That ever the variegated birds reveal
The spots or stripes peculiar to their kind,
Spring after spring: thus surely all that is
Must be composed of matter immutable.

LUCRETIUS, *De Rerum Natura*
Translated by WILLIAM ELLERY LEONARD

7 a.m., April, the Pennine hills

In chapter 4 I left Darwin's entire theory of sexual selection by female choice in tatters, dismissed and degraded by Julian Huxley for the second time, as it had been by Alfred Wallace a generation before. Huxley's unsatisfactory 1938 papers on the topic proved highly influential, being cited throughout the 1950s and 1960s as the best, last word on the topic. There might be some female choice in the Peafowl, the Ruff or the Black Grouse, he conceded grudgingly, but the vast majority of colourful male birds were the product of some form of natural selection for a practical purpose: intimidation of other males or just simple species recognition. Darwin was wrong to call sexual selection a special or a widespread force in evolution. You could forget all about his notion of an aesthetic sense in female birds, let alone one shared with Sir Joshua Reynolds, being behind the development of beautiful male birds. And if sexual selection did happen, some scientists were saying by the 1960s, it was a bad thing. The Austrian pioneer of ethology Konrad Lorenz wrote in 1963 that 'the evolution of the Argus Pheasant has run itself into a blind alley. The males continue to compete in producing the largest wing feathers, and these birds will never reach a sensible solution and "decide" to stop all this nonsense at once.' The fools.

But there were stirrings of rebellion once more against this lazy consensus. Peter O'Donald, who had been Ronald Fisher's last doctoral student at Cambridge University, tried to cast off the Huxleyan confusion that surrounded Fisher's runaway hypothesis. O'Donald was studying a seabird in the Shetland Islands called the Arctic Skua and finding an interesting mate-choice effect. Arctic Skuas come in three forms, dark, light and intermediate. The light birds have a creamy white breast and belly where the dark ones are dark grey-brown all over. There was little to choose between them in when they laid eggs or how successfully they bred. But O'Donald found that when a male was looking for a new mate, either at the start of adult life or after bereavement, it took him about six days to find one if he was a dark male, about eight days for an intermediate male and about eleven days for a pale male. There was, in other words, a clear female preference for pairing up with dark, handsome strangers – and in a monogamous species, not a lekker. Could female preference and agency really be so easily dismissed?

In 1962 O'Donald published the first mathematical models of Fisher's runaway mechanism, demonstrating that a mutation that enhanced a female preference and that was also linked to a male trait used in display would be picked out by sexual selection: 'Thus we should expect to find that one or more supergenes control male display and female response to it.' But O'Donald's mathematics remained obscure and few took notice.

In 1972 a book was published to mark the centenary of the publication of the *Descent of Man*, with eleven chapters by learned evolutionary biologists. Darwin was given high marks by various authors for his speculations about human origins among African apes, but his theory of sexual selection garnered rather more lukewarm enthusiasm from the examiners. The chapters on the topic went over old ground and added little new, hewing partly to Huxley's downbeat line, though Ernst Mayr did at least note in passing that Peter O'Donald had shown the Fisher effect to be feasible. However, one chapter in the book proved ground-breaking, written by a young Harvard scientist who had recently switched from social science to biology. His name was Robert Trivers and his idea was so simple it would probably have left Darwin cursing: 'I wish I had thought of that.'

Trivers burst upon the scene of evolutionary science with a string of insightful intuitions and the mathematical proofs to go with them, earning him the accolade 'one of the great thinkers in the history of Western thought', from no less a great thinker than Steven Pinker. It was Trivers who inspired much of the revival of Darwinism and its application to human psychology in the later twentieth century. Widely regarded by his acquaintances (including me) as a near-genius, Trivers is also a controversial character, having joined the Black Panthers, fallen out with most of his university employers, suffered from bipolar disorder, and while living in Jamaica smoked a lot of cannabis, got into plenty of trouble and been arrested on several occasions.

Some time around 1970 Trivers was looking out of a bathroom window in his third-floor apartment in Cambridge, Massachusetts, watching two pairs of pigeons breeding on a nearby roof. Male and female pigeons are indistinguishable in appearance (though not in behaviour) and breed in monogamous pairs that share parental duties.

But what intrigued Trivers was that despite this apparent equality there was a marked difference in their behaviour. The males jealously guarded the females, not vice versa. The males were aggressive to other members of their sex, which the females were not. The males spent a lot of time trying to seduce other females, which showed little interest; the females did not spend time trying to seduce other males.

Why the double standard? Why is the male usually the one asking for sex and the female the one to grant it only when it suits her, even in monomorphic pigeons? Trivers's answer was deceptively simple. Whichever sex invests the least time, risk and energy in reproduction does the proposing, and whichever sex invests the most does the disposing. The discrepancy between the sexes begins with large eggs versus small sperm but goes much further than that. In people – and most mammals – because of internal gestation and female lactation, the female devotes a vastly greater flow of energy and chunk of time – years in our case – to bearing and suckling a baby. Not only does this dwarf the investment of even the most dedicated male; it also leaves her vulnerable to desertion by the male, or being left 'holding the baby'. Females can either try to get some guarantee of male fidelity, provision or assistance before they choose to mate – as they often do in human societies – or resign themselves to being fought over by brutally power- ful, but unhelpful males, as they usually do in Elephant Seal societies. Trivers suggested that the less males contribute, the more they will compete over females and the valuable resource they represent. The more males invest, the more selective they too will be.

In Black Grouse the investment is starkly asymmetric. The males supply about two seconds of parental investment, called copulation, the females about four weeks of incubation and three months of paren- tal care. According to Trivers, therefore, the females should be choosy and the males eager. Tick. The females should be dull-coloured, the males bright and ornamented. Tick. In the Red Grouse where the father guards his mate before and after she lays her eggs and accompanies his young family for weeks if not months, there should be much less striking sexual dimorphism. Tick. Suddenly we have a clear and useful theory for deciding why males are (usually) the more gorgeous sex, and how much more gorgeous we should expect them to be.

But birds are different from mammals in a key respect. Though female birds are the ones that lay eggs, males can incubate them and neither sex breastfeeds them (pigeons being a rare exception where both sexes produce a sort of milk from their throat, which may explain why many pigeon species have little sexual dimorphism). So it is possible for male birds to invest more than females and if they do it will be females that fight over males – according to Trivers's prediction. Nearly half a century ago I spent half a summer camped with three colleagues, David Perkins, Charles Gillow and James Ogilvie, on an island in the high Arctic watching a bird in which this is the case: females are more brightly coloured and pugnacious than males. There are maybe three dozen species of bird in which this is the case but no mammals. They are exceptions that allow us to test Trivers's very simple rule.

'The female instead of the male takes on a gay and conspicuous nuptial plumage'

The bird I was watching is a fabulously unusual species. The size of a Starling, it lives on the ocean waves for nine months of the year around the capes of Good Hope and Horn, bobbing about on giant rollers in especially stormy seas, often following whales to catch the plankton stirred up by the leviathans. Then for three months of summer it turns up nine thousand miles to the north in bogs on the tundra near the shores of the Arctic Ocean. In both places it meets very few people and as a result it has none of the wariness of most birds. I have stroked them as they sat on their nest and watched them fight almost between my feet. Known in Britain as the Grey Phalarope and in America as the

Red Phalarope, the confusion of the name results from its different plumage in autumn and winter, when it turns up around the British coast, versus spring and summer, when it breeds on marshes in the high Arctic of Canada and Alaska. It's a small sandpiper but with partly webbed feet and a preference for swimming over wading. On little pools in a marshy valley, overlooked by glaciers, as rafts of ice melted with a tinkling sound on the nearby lakes in the midnight sun, we watched this tiny creature begin a frenzied, brief breeding season fuelled by a rich supply of midge larvae and small crustaceans. (We probably could not do this study today because of the increase of Polar Bear numbers in the area.)

At first sight there is nothing odd about the phalarope's courtship. Bold, brick-red males (it seems) with blackish backs, black and white heads and yellow bills fight over duller, paler, brown-backed, brown-headed females (it seems), flying after them and pursuing them on the water with puffed-up feathers, in what a Russian scientist, A.A. Kistchinski, had described a few years before as a mobile lek. Then suddenly mating happens and it's the dull bird that mounts the bright one. This is a female lek, not a male one. As Alfred Russel Wallace wrote in 1870 of this very species:

> Yet it is undoubtedly the fact, that in the best known cases in which the female bird is more conspicuously coloured than the male, it is either positively ascertained that the latter performs the duties of incubation, or there are good reasons for believing such to be the case. The most satisfactory example is that of the Gray Phalarope (Phalaropus fulicarius), the sexes of which are alike in winter, while in summer the female instead of the male takes on a gay and conspicuous nuptial plumage; but the male performs the duties of incubation, sitting upon the eggs, which are laid upon the bare ground.

Much the same is true of the two other species of phalarope, the Red-necked and Wilson's. Females are larger, more brightly coloured and more aggressive.

Here Wallace gives a clue to what is going on. The male, for some reason (I will return to this), is the one that invests the most time and effort in parental care, so it is the male that is the scarce resource in the market at any one time, fought over by the sex that has more time on its hands. To put it in technical jargon, the 'operational sex ratio' favours females in this species: at any one time there are more spare females hanging about than spare males. As Trivers spotted, this is a general rule in the animal world. Whichever sex devotes most time and energy to raising the young is small, dull and choosy; whichever sex devotes least time and energy is large, colourful and sexually forward.

Red-necked Phalaropes: male on the left, female on the right.

In the Arctic that summer we watched females lead males in a strangely ceremonious fashion into the sedges around a pond, bowing low to the ground in certain spots as he watched. If he moved away she followed and tried again. Eventually he liked one of the spots she had chosen and fashioned it into a cosy little nest. The ceremony was designed – we thought – to overcome the unusual problem that although she will lay the eggs it is he who will be spending the next two weeks sitting on them.

For obscure historical reasons, phalaropes, and a bunch of other birds such as jacanas, dotterels, painted snipe, tinamous and button quails, have got into the habit of the males sitting on the eggs and guarding the chicks, meaning that they end up investing more time and energy than females. In all these species females are more gorgeous. A clue to how this possibly came about lies in the 'double-clutching' habits of some other Arctic sandpipers, in which the female lays two clutches of eggs, one for her mate to sit on and the second for her to sit on. Once that habit was established it would be quite easy for a cunning female to lay her mate's clutch and then find another male, lay him a clutch, then look for a third in a form of serial polyandry. She somehow never gets around to laying her own clutch, each of her successive mates being landed with another 'first' clutch. Each is left 'holding the baby'.

If this tactic was successful, she would leave more offspring than rival females, thus founding a dynasty of duped but dutiful males and serially unfaithful females, and the habit would spread. Soon, the more successful females would be the ones that fought over spare males, driving away their rivals and actively courting every male. Extra testosterone would help them to be both pugnacious and brightly coloured – research has shown this to be the case. Eventually the phalarope species that I watched has come to be dominated by red females that attempt to seduce brown males whose services they need to raise their young. Today each female phalarope lays a clutch of four eggs for a male and

Male Dotterel brooding eggs in Scotland.

then deserts her mate, spending a week or two trying to seduce other males before heading back to sea long before her dutiful spouse. Incidentally, if this history is correct and the sex-role switch began with double-clutching, it means that the species is now half as productive as it once was, rendering it more vulnerable to extinction.

The male phalaropes indulged in one behaviour that I struggled to explain. A male would sometimes leap off the nest and fly over to attack a nearby female, repeatedly hitting her with his feet, twittering furiously, till she flew off. His anger was clear and seemed wholly out of character with his meek behaviour while being courted by more aggressive females the week before. This was the 'driving flight', a term coined by Kistchinski who had seen the same thing a few years before in eastern Siberia. He thought it was directed at the male's mate. I disagreed, observing that it was not directed at the mate but at strange females. It appeared to be a case of defending the nest from intrusion by a different female. For instance, one afternoon I watched a male leave his nest to feed. A female turned up and flirted with him, puffing herself up into the courtship posture and doing little bursts of hovering known as 'rattling'. He then began to return to his nest and she followed him. He immediately attacked her (my notes say, 'He tells her to bugger off') but this attracted another female, which he also had to attack. His own mate, meanwhile, who I recognised from the shape of the white patch on her head, was off on another pond harassing another male.

A couple of years later, while watching a television programme, I suddenly realised why these males had been so determined to drive away strange females who got near their nests. The documentary, about birds, showed a male jacana, a tropical bird that has long toes so it can trot across lily pads, leaving his nest to attack a larger female in almost exactly the same way. The reason soon became clear: she was intent on breaking his eggs and indeed managed to do so when he was distracted. Infanticide is a well-known tactic employed by individuals of certain species of mammal, such as lions, langur monkeys and Gorillas, to free up potential mates for fresh child-bearing duties. In those mammals it is the male that does the killing. In jacanas and – I am all but certain – phalaropes, it is the females. By destroying another female's eggs, she hopes to free a male to sit on her next clutch. Indeed, looking back at my data, it may explain why seven of the fifteen nests we found were destroyed without explanation. At the time I blamed the Arctic Foxes and skuas that very occasionally patrolled the marsh; I now suspect that female phalaropes were the culprits.

The calculus is clear: animals fight over valuable but limited resources. In phalaropes that limited resource is incubating males; in Black Grouse it's incubating females. The sex that invests least time and effort in bringing up the young fights over the sex that invests the most

– and also ends up most beautified. At some point in the ancestry of the Black Grouse, the most successful females were ones that chose males for qualities other than fidelity and domesticity. They chose them for their skill at flamboyant display instead. The male Black Grouse invests almost nothing in his offspring: just a package of sperm delivered in seconds. In theory he could supply hundreds of females. No wonder they compete so fiercely and for so long.

'A glaring non-sequitur'

Trivers's theory made little sense except in the light of Darwin's theory of sexual selection. Whether it takes the form of competition among males, or choice by females, the different opportunities open to each sex predict how colourful each sex would be. Thus it breathed new life into one aspect of sexual selection.

In 1980 the soft Huxleyan consensus – that Darwin's idea of female choice was a marginal and minor phenomenon – received another challenge, as it had from Fisher fifty years before and O'Donald in 1962. Like them, the University of Chicago's Russell Lande was a pioneer of new mathematical techniques in genetics for calculating the effect of natural selection on genetic traits. In 1980 he turned his attention to sexual selection and built an elegant and detailed mathematical model in which he showed very clearly that given reasonable assumptions, the runaway effect would indeed work. In dry technical terms he spelled out the conclusion: 'A male character that is under stabilising natural selection towards a phenotype that is optimal with respect to survival may evolve a markedly suboptimal phenotype through sexual selection acting through mating success.' Translated into simpler English: an ancestral Black Grouse with normal tail feathers that were being continually honed by natural selection to be neither too long nor too short might suddenly find itself evolving instead into a bird with a lyre-shaped tail that positively hindered its survival in normal life, if females suddenly develop a coordinated preference for displays of lyre-shaped tails.

A leading evolutionary theorist, John Maynard Smith, drew the scientific world's attention to Lande's paper and it became an instant classic. Suddenly, for the third time, female choice as an engine of evolutionary change was in fashion. Lande drew attention especially to the fact that his model could explain why related ornamented species could have very different males even while their females looked much the same: the process could run away in any, arbitrary direction in an isolated population.

Lande's model was soon followed by another done by Mark Kirkpatrick of the University of Washington in Seattle that showed the Fisher effect to work with even simpler assumptions. Both Lande and Kirkpatrick proved Fisher's hunch correct that female preference and male ornament would evolve together, reinforcing each other. Despite making radically different assumptions about genetic mechanisms, they both showed that the sexy-son effect worked and worked fast. That is to say, with a modicum of heritability of both taste and trait, the two would run away together into fantastical forms. Astonishingly, Kirkpatrick concluded that a chance mutation in a female that gave her a random preference for some particular male trait could carry that trait with it as the mutation spread 'even if the male trait is nearly lethal'. As a parallel, consider that the breeders of German Shepherd dogs in recent years have become so obsessed with exhibiting them at shows with sloping backs that the breed is now plagued with hip dysplasia. Likewise, female choice can burden males with dangerous features. There is little doubt that a Peacock is disadvantaged by its tail: I once caught one by the legs as it displayed because it could not see me coming from behind. So the possibility emerges that females may be unwittingly breeding into males some severe handicaps to survival.

Back in Cambridge, O'Donald had been thinking that Fisher had solved Darwin's problem of why females should prefer male ornaments at all. But by the 1980s he had come to realise that Fisher's argument contained 'a glaring non-sequitur'. Suppose a female in a species of ancestral Peafowl likes a male with eye spots on its rump. Why would it prefer a male with bigger eye spots? It does not necessarily follow that a feature preferred by females would be even more preferred if even more extreme. Fortunately for the theory, it does seem to be generally true that birds – and people – like 'supernormal stimuli'. The classic example was a study by Niko Tinbergen showing that if you present a bird with an enlarged egg, or one painted in a more saturated version of the normal colour for that species, it tries to brood the fake egg in preference to its own. So O'Donald's worry could probably be laid to rest. Presented with an unrealistically enhanced and exaggerated version of a Peacock, a Peahen would be only too happy to mate.

If sexual selection by female choice was to persuade many people, however, it would have to go beyond theoretical proofs with equations, however elaborate. Neither Darwin's and Fisher's verbal arguments, nor O'Donald's, Lande's and Kirkpatrick's mathematical models would ever be enough to convince sceptics that they had discovered a real phenomenon, capable of generating beauty and extravagance in the real world. What was needed was an experiment.

Step forward Malte Andersson. As a young zoologist at Gothenburg University in the 1970s, Andersson studied a bird called the Long-tailed Skua and read O'Donald's work on sexual preference in its larger cousin, the Arctic Skua. He became convinced that the exceptionally long tail of the smaller species, which switches every spring from stealing fish off seabirds to eating lemmings on the tundra, was not needed for the birds to recognise their own species, as had been argued, Huxley-fashion. Instead the long tail streamers were probably an exaggerated ornament used in display – in this case by both sexes. He published a theoretical paper on sexual selection arguing that if orna-

The Long-tailed Skua feeds on lemmings in the summer.

ments are sexually selected they will probably show gradual development over several seasons – as is the case in Peafowl, for example, where males do not grow full trains till they are three years old. Andersson watched Peacocks displaying in the wild on a visit to Sri Lanka and began to think about deliberately manipulating their trains to make supernormal tails to see if it made a difference to their ability to attract mates. But working with wary Peafowl in a forest with wild elephants turned out not to be a practical project.

On a visit to Kenya, Andersson had a better idea. In grasslands grazed by herds of Kikuyu cattle in the Kenya highlands, he saw Long-tailed Widowbirds flying over the long grass. The female widowbirds are brown, sparrow-sized birds that nest in domed nests built into the stems of tall grass. The males are black, with bright red shoulders and vastly long, thick tails, far longer than their bodies, hanging half a metre below them when they perch on fenceposts. The males set up territories in the grasslands and display by flying slowly about with their huge tails bent downwards but trailing backwards in a curved and spread shape, like the bow of a ship. It is pretty obvious to the neutral observer that what is going on is that the male is shouting to females, 'Look at my tail, please mate with me!' Sure enough, the males are polygamous, a successful male bird persuading up to five females to nest in his territory.

Andersson found a spot on the Kinangop Plateau above the escarpment of the Rift Valley in central Kenya, where the birds were plentiful. With the help of two local assistants, Uno Unger then Kuria Mwaniki, he then did something so simple yet so clever that it became a landmark in the history of evolutionary studies. He trapped thirty-six males in clap-net traps and divided them into four groups. He shortened the tails of nine birds by cutting them; lengthened the tails of another nine by attaching the cut feathers from the first group on to this group's tails with superglue; left nine alone except for rings on their legs; and cut the tails of the final nine but glued the feathers back on without changing the length. These last two groups were 'controls' to check whether the treatments themselves changed anything. The birds were then released back on to their territories.

He had already counted how many nests the females had built in each male's territory – as far as he could tell females nested in the territory of the male they mated – and they were roughly equal on average between the four groups. He found the nests by walking systematically through the grass, a time-consuming but not difficult task, repeating the search once a week till the end of the breeding season. There seemed to be no correlation between the number of nests and the length of the male's tail at this stage, despite 9 per cent variation in tail length. Now, however, the nine males equipped with supernormal stimuli in the shape of elongated tails attracted significantly more females to nest in their territories: an average of 1.9 nests per territory. The males with shortened tails had significantly fewer nests: an average of 0.4 nests. The controls attracted only slightly more than the short-tailed birds, so the striking result was that the nine elongated males got more nests

The male Long-tailed Widowbird displays while the female watches.
Photos by Adam Scott-Kennedy.

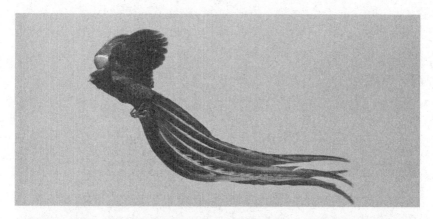

(17) than the 27 others (16). The successful males had not expanded their territories, suggesting that the longer tails did not help intimidate other males. Nor did they increase the rate of displaying: they actually displayed slightly less often whereas the males with shortened tails displayed more often, presumably as they grew increasingly desperate to attract mates.

By far the most plausible explanation for Andersson's data was that elongating the tail made a male more attractive to females. Here was clear vindication of Fisher's hypothesis, that a male ornament has been exaggerated beyond any sensible size by the action of increasingly choosy females. Here too was the answer to the non-sequitur O'Donald had found in Fisher's argument: the females did indeed appreciate tails even longer than nature provided, the supernormal stimuli.

In 2016 Andersson reflected on how the experiment had changed his life. After weeks in 'a beautiful rural part of the Kenya highlands, with cool nights and hot sunny days, helpful and friendly Kikuyu farmers, and the Nyandarua Mountains as a magnificent backdrop under a clear sky', he wrote up his paper, published it in *Nature* magazine and quickly became sexual selection's star player, commissioned to write a huge textbook on the now sexy topic, which with revision after revision ate up much of his time over the next few years.

'Indirect fitness gains via increased attractiveness of sons'

Throughout the 1990s many different lekking species were put under the microscope to try to crack the puzzle of how sexual selection really works. One lovely case was the stalk-eyed fly, *Cyrtodiopsis dalmanni*, which lives along streams in Malaysia. The males have bizarre faces with the eyes out on thin stalks at right angles to the head. In the evening, they gather on dangling plant root hairs along the banks of streams and fight each other. The females like to roost on such roots, so the males await their arrival and court them. Mating takes place at dawn. Cunning scientists come along with nylon stockings and capture all the flies on one root so they can measure what is going on. It seems that the males with the widest-spaced eyes win the most fights and attract larger clusters of females. The scientists then raised the flies in the lab and saw females preferring to roost with longer-eyespan males. It seems that the cause of the bizarre eye-stalks is female preference. But once again the scientists were unable to discriminate between good-genes and sexy-son in their attempts to determine why the females were choosy.

Next, by forcing one line of females to choose the longest-eyespan males and forcing another line of females to choose the shortest-eyespan males for thirteen generations, they created lineages of females with consistent, inherited preferences. Sure enough, the first line preferred long eyespan, the second line preferred short eyespan. That at least supports one of Fisher's key predictions: that the trait and the preference would come to be genetically correlated as they evolve together.

Bizarrely, this species introduced an entirely new theory into the debate. Twenty years after the field studies, Andrew Pomiankowski and his colleagues at University College London found that the stalk-eyed flies carried a special genetic mutation that biased the sex ratio of the offspring. Males with smaller eyespan had more female and fewer male young. So perhaps, by mating with long-eyespan males they were avoiding this gene that biased the sex ratio. Not avoiding a parasite as such, but a sort of genetic parasite.

So how would you go about testing whether Black Grouse plumage was created by a runaway, sexy-son effect? How can I find out what is (unconsciously) in a greyhen's mind when she chooses Black Spot rather than Wonky Tail at my lek: is it the need to have sexually attractive sons or the wish to have healthy offspring? Well, ideally, you would present females with a selected set of males, observe their choices, then measure the survival and fecundity of their sons and daughters and also measure the attractiveness of their sons by presenting those sons to a new generation of females in a controlled experiment. That's a tall order but amazingly and fortunately it has been done – though in flies rather than birds.

A species of small, blood-sucking sand fly found in Brazil, called *Lutzomyia longipalpis*, is a lekker. At night males – which do not suck blood – gather around chicken sheds or the bodies of sleeping mammals and birds to await the arrival of females, which do suck blood. The males defend small territories, wing-fluttering and performing hopping displays at approaching females, while emitting pheromones and 'singing' – again by fluttering their wings. Just like miniature Black Grouse. On arrival the females drink some blood and then sample several males before choosing one as a mate, after which they go away to lay a brood of fifty or so eggs. Like Black Grouse, the female flies are in charge of deciding whether mating happens; they reject males by 'depressing their abdomens or moving away' from courting males. Unlike Black Grouse, an entire lek can be recreated inside a laboratory and an entire breeding cycle observed within weeks.

In order to work out what was going on, some years ago, a Cambridge scientist I have known for a good while, Andrew Balmford, supervised a student, Theresa Jones, to establish a colony of the flies in a laboratory in Brazil in such a way that she could measure the impact of mate choice on the next generation. She took ten virgin, blood-fed females and one by one introduced them in turn into a caged lek with five males in it. In each case, the females generally agreed which of the five males was attractive and this male fly got most of the matings. Jones repeated this twenty-five times with different flies each time, then took the least attractive males from each trial and grouped them into six new five-male leks. Ten new virgin females were given the chance to choose among these losers. She then followed the females –

186 in all – to see: first, how long they themselves lived and how many eggs they laid; second, how well their eggs and larvae survived, how well their adult offspring did and how well their adult daughters bred; and third, how attractive their sons were to a new generation of females.

If the purpose of lekking was to give the females themselves a direct benefit, such as sorting out the most fertile or least infectious males for them, then the females that chose the best male should have survived better or be more fecund. They did not. If the purpose of lekking was to seek out good genes for the next generation, then those offspring should show higher survival or fecundity. They did not. If the purpose of lekking was to breed sons that will be chosen by females at future leks, then the females that chose attractive males should have had sons that are more attractive than the sons of females who were forced to choose unattractive males. They were – strongly so.

In other words, at long last, in a lekking species, somebody found a way to pit the two main hypotheses – Darwin versus Wallace, Fisher versus Huxley, sexy-son versus good-genes – against each other. And Darwin-Fisher-sexy-son won, hands down. Of course, this is just one experiment and it is possible that it missed some obscure good-genes benefits such as disease resistance, but Jones and Balmford and their colleagues argued that they would surely have seen some good-genes effect if it was there. As they wrote in their paper, published in 1998, other studies claiming to show good-genes effects had not even looked for Fisherian benefits. In effect, they suggested that the benefits of having attractive offspring might be more valuable than the benefits of having healthy offspring – delivering what they called 'indirect' fitness in subsequent generations.

7.

Curlew Chorus

Teach us, Sprite or Bird,
What sweet thoughts are thine: I have never heard
Praise of love or wine
That panted forth a flood of rapture so divine.

PERCY BYSSHE SHELLEY, *Ode to a Skylark*

7.30 a.m., April, the Pennine hills

The females have left and the lek is a bit more peaceful. The males are still here but the displays now come in sporadic and half-hearted bursts. No longer distracted by the action, I suddenly notice the background music. The dawn chorus is continuous, overlapping and glorious. The Oystercatchers are the loudest, the Peewits (what southerners call Lapwings) seem the most joyful, the Redshanks sound the most urgent, the Golden Plovers have the purest tones, the Skylarks sing the most complex songs. But it is the Curlew that catches the ear. Anthropomorphism be damned: I defy anybody not to describe the throbbing crescendo of the Curlew's song as 'joyous'. It's somehow just perfect for the place, a theme tune for the landscape. As the Scottish farmer and nature writer Patrick Laurie put it in his book *Native*, 'No other wild bird has that power to convey a sense of place through song.

It's a grasping bellyroll of belonging in the space between rough grass and tall skies, and you never forget it.'

The calls of Skylark, Golden Plover, Redshank, Peewit, Cuckoo and all the others that I hear this morning are also thrilling to the human ear: excited, lively, lyrical, pleasing and just plain happy. Above all they are beautiful. As I listen to their calls, I realise that this is odd. I last shared a common ancestor with a Curlew or a Skylark about four hundred million years ago, when that ancestor was a lumbering Devonian reptile that presumably grunted – if it called at all – in its swampy home. Can there really be a universal standard of beauty that is shared by such different beasts as me (a descendant of small, nocturnal Jurassic shrew-like creatures) and the Curlew, a long-beaked descendant of mid-size, diurnal Jurassic dinosaurs? How can it be that

The Curlew's song is a theme tune for the moorland landscape.

the aesthetic sense of the bird today coincides partly with mine, that we both have, in Darwin's word, a similar 'taste for the beautiful'? By contrast, the Curlew's anxious whinnying call in June when I walk too near its chicks is not beautiful at all: just unhappy. Likewise, the Peewit, Redshank, Oystercatcher and Golden Plover also sound deeply worried – to the human ear – when they have chicks on these moors.

The Black Grouse female undoubtedly finds the black, blue, white and red of the male in full display aesthetically pleasing, and so do I. I can think of few male bird displays, or songs, that are ugly. Show somebody a film of a Peacock displaying, or an Argus Pheasant or a bird of paradise, or play them a Nightingale or a Curlew, and they will exclaim not 'How weird' or 'How dull', but 'How lovely!'

I'm back to the point that so puzzled Darwin. You can just about devise a theory to explain why male ornaments – and songs – became conspicuous and exaggerated, but how or why did they become so beautiful to the human eye and ear? It's hard not to think that birds too must find them so, that the aesthetic sense is one we somehow share with Peahens and Curlews. In the *Descent of Man*, Darwin wrote: 'On the whole, birds appear to be the most aesthetic of all animals, excepting of course man, and they have nearly the same taste for the beautiful as we have.' The song of the Curlew, now reverberating around the moor as I sit here, is a thrill to the human soul: it builds to a crescendo of exhilaration as deliberately and skilfully as a passage from Beethoven. Like Ludwig's version, the Curlew's ode to joy begins with a slow repetition of a portentous single chord (cellos) as it starts to rise in frequency and speed of repetition (violins). The chords come quicker, higher in pitch and louder (trumpets) till they suddenly break, like a wave on the beach, into a rich bubbling trill (chorus). At this point the bird is often a hundred feet high in the sky almost hovering on rapid wingbeats and as it stalls at the top of its arc and glides back towards the ground the song begins to slow down again and drop in pitch till it has faded away, as does Beethoven's version. It cannot be a coincidence that composers and Curlews use the same pattern of crescendo and diminuendo to appeal to their audiences.

The standard of beauty that bird song, and bird feathers, share with people would include a preference for pure notes, rather than harsh squawks or clicks, harmonies rather than disharmonies, regular, rhythmic phrasing rather than a jumble, and in the visual world, pure hues, bright luminance and elegant shapes. There may be a hint here as to why my aesthetic sense and the Curlew's and Peahen's are so similar. Purity of note or hue is harder to produce, more statistically improbable and more unusual than broad-spectrum noises or broad-spectrum browns. In the natural world, bright colours and pure notes do not occur by accident; they have to be created deliberately.

Yet surely we human beings only discern a tiny fraction of the complexity and the thrill in a bird's song. Just as we must seem bizarrely oblivious and anosmic to a dog in not paying attention to scents, or blind to a Starling in not seeing ultraviolet patterns, so we must be deaf to many of the intricacies of meaning in a Curlew chorus. As the great African scientist-artist Jonathan Kingdon put it in his book *Origin Africa*, 'A flock of piapiacs or a covey of crested guineafowl have vocabularies, complexities of syncopation, burdens of meaning to match a human composer's range of composition, a sensitivity to pitch that would please Art Tatum.'

Curlews call to each other all year but sing mainly in the spring. Their song is their equivalent of the Black Grouse's display. (The Nightingale's song has often been called an acoustic version of the Peacock's train.) Unlike Black Grouse, Curlews are monogamous, forming pairs that occupy partly exclusive nesting territories, and they remain wedded for years. Both sexes sing. In April, as new pair bonds are forming, old ones reforming after the migration to the breeding grounds and territory boundaries being established, the songs of Curlews here are continuous and overlapping. I sometimes challenge friends to count to ten without hearing a Curlew singing during the early morning at this time of year. They always fail.

'Dear enemies'

Superficially it is puzzling that birds should continue to indulge in conspicuous display throughout the spring if they are faithfully wed to one mate. Sexual selection was supposed to be confined to polygamous birds, according to most of the biologists who thought about the topic in the twentieth century. So why do Curlews and Skylarks sing throughout the spring? Why don't they sing till they get a mate, then shut up?

The explanation for this came from genetics. In the 1990s, after Sir Alec Jeffreys had stumbled on the fact that every individual has unique patterns in their genomes, DNA fingerprinting became available and biologists started to discover something shocking about birds: many of them were not their ostensible father's offspring. Birds that lived in monogamous pairs and shared the duties of raising the chicks were doing really quite a lot of 'extra-pair copulation'. Hence all that singing. Even as his mate is laying or incubating eggs, a male Curlew or Skylark is hard at work advertising his quality to the neighbouring females in case they fancy a secret liaison. And he is sometimes successful. In one study of Skylarks in south-west England, 171 chicks from 52 broods were genotyped and 35 chicks (20 per cent) in 14 different broods turned out to be fathered by males other than the ones that were feeding them. They all matched their mothers so it was not a case of females laying eggs in each other's nests. Similar studies have not been done on Curlews, as far as I can find, but in two different populations of Curlews in Finland, one of which was denser than the other, the birds in the dense population did more singing and more mating, much of it in response to intrusion into the territory by another male,

Skylarks have distinctive dialects.

suggesting that there is a good deal of need to sing if you are a male Curlew, either to seduce the neighbour's mate or to persuade your own mate not to seduce the neighbour's partner.

Curlew song is loud and lovely but not very varied. More complex than a Cuckoo's song – which also delights my ear today – but just as stylised. Each male sings the same melody, varying only in the length of the build-up and the extent and power of the bubbling crescendo at the climax. Skylark song, on the other hand, is almost infinitely variable. The phrases in it are recognisably those of the species but they change continuously and are mixed in ordering so as to produce a virtuoso performance of staggering, maybe infinite variety. Scientist Thierry Aubin and his colleagues in France have discovered that song plays a sophisticated role in Skylark society. The birds breed in loose groups of pairs and each group turns out to have a unique signature of song, in which the order and pattern of the phrases is distinct. Aubin calls it a micro-dialect. When he plays back to a Skylark a song from a different group, the bird responds much more aggressively than to one from his own group. Likewise if he jumbles the phrases around, the bird thinks it is hearing a different dialect from a different group, and approaches aggressively. In short, Skylarks know their neighbours, but see them as lesser threats – 'dear enemies' – than strangers. This is a bit like the Black Grouse on the lek: neighbours on the lek are enemies, yes, but well-known ones with whom a truce has long been declared, unlike intruders from nowhere that sometimes turn up and are violently driven off.

Cuckoo chick in a Meadow Pipit nest near the lek.

Just as with the displays of Peacocks, there was a general reluctance among twentieth-century biologists to accept Darwin's idea that male birds sing to attract mates rather than to deter rivals. But while some birds such as Robins seem to use song as a form of territorial defence against other males, many birds such as Sedge Warblers abruptly cease singing once they have a mate; yet others such as Great Reed Warblers sing more briefly and simply once they have a first mate, but sometimes go on to acquire several mates. So the suggestion that the point of song is often to attract females is and always has been highly plausible.

It was not until the 1980s that scientists first did experiments to prove the simple point. Dag Eriksson and Lars Wallin put out nest boxes for Pied and Collared Flycatchers in Sweden. On each box was a dummy male bird, but on half the boxes there was also a loudspeaker playing bird song. And each box was specially designed to trap any female that landed on the perch at the front. The traps caught ten females, nine of which were at boxes playing song. Other similar experiments followed. On a Sage Grouse lek in America, playing recordings of the calls of males attracted more females to the lek.

Zoologist Clive Catchpole did a series of experiments with Sedge Warblers and Great Reed Warblers starting in the 1980s that gradually untangled what was going on. Both species sing highly complex songs with a huge repertoire of phrases. A Sedge Warbler has a song list of about seventy-five different types of phrase and will generally sing for twenty seconds during which he will voice about three hundred different elements. Females would work their way through several territories in a reedbed, apparently sampling the songs of different males before returning to one with a song they liked. And male Great Reed Warblers with more complex songs attracted more females (the species is often polygamous) and produced more young. So it is fairly obvious how complex bird song came to evolve: females chose it.

Still, pull back from the scientific details and marvel at the strangeness of this planet. Here's a small brown bird impressing a female by uttering a rapid succession of squeaks, squawks and trills, many of which it picked up on its winter holiday in Africa by listening to the local birds singing there. Why? Being good at remembering and imitating as many songs as possible cannot surely make a male a good father, except in the general sense of proving him to be healthy – and there are surely other ways for him to prove his general health. We are back once again to the argument from incredulity that entrapped old Wallace. It's just hard to believe that an individual female would mate with a male, committing her genes to combine with his, committing weeks of sterling efforts at parental care, committing her one chance in a year to leave behind a genetic legacy, committing all this on the whim of liking a complicated aria. It is so astonishingly random. Which is perhaps why Fisher's explanation is so appealing. It says that anything can become an exaggerated sexual signal – any colour in the plumage, any elongated type of feather, any type of song, any aspect of virtuosity in singing.

In striving to understand why a female Sedge Warbler might fall in love with a good mimic of African cisticolas, I can appeal to a more familiar example: human beings. People have huge vocabularies, regularly knowing sixty thousand words each, and learning them with astonishing rapidity in childhood, by imitation. This is far more than is really needed for an efficient communication system and our use of language is anything but efficient. We deploy these many words in many ways, but prominent among them is courtship. At mating age, human beings are all to some extent Cyrano de Bergeracs or Scheherazades, trying to impress members of the opposite sex with songs, stories, rhyme and repartee. Verbal dexterity tends to impress and even seduce. Long phone chats, long letters, long social-media exchanges and long face-to-face conversations are central to courtship, and people end up married to mates who have similar-sized vocabularies.

Nor is it just the complexity of the song that counts. In Canaries, it seems there are certain 'sexy syllables' in the song that really turn the females on. Those that heard artificial songs with extra helpings of such sexy syllables produced more sex hormones in their blood, laid larger eggs and deposited more testosterone in the yolk of the eggs, as if intent on generating more masculine offspring. There is even a study that finds that playing the sexy syllables to captive birds switches on a particular gene called ZENK in the female brain – more so than playing a tone or random sounds.

In American song sparrows, captive females responded more strongly to songs with a larger repertoire of song types, while out in the field males with larger repertoires were more likely to acquire mates. Likewise in Catchpole's Sedge Warblers, females displayed most to recordings of larger repertoires of Sedge Warbler song. Males with larger repertoires had larger territories and were less likely to be infected with blood parasites. In Great Reed Warblers, the males that achieved 'extra-pair copulations' had larger repertoires than the males they cuckolded, and the young they fathered were more likely to survive to breed in future years. Also, experiments found that deliberately stressing young male birds by half-starving them or feeding them stress hormones results in adults with shorter, simpler songs.

All of which suggests that – in contrast to Fisher – there may be method in the madness of song. Females are choosing to mate with good singers and that way they probably get superior genes for their offspring. Some species such as Curlews stick to simple, repetitive songs and presumably it is the vigour, frequency and purity of singing that marks out the best males. Others such as Skylarks have gone in for virtuosity to the point that they fill their songs with new phrases, many copied from other bird species, and presumably in such species it is the diversity of the repertoire that most impresses females. But let's not forget that by choosing the most diverse repertoires they also would automatically get sexy sons. Nothing in these intricate studies of bird song can so far clearly distinguish Fisher's explanation for the Curlew, Sedge Warbler and the Nightingale from Wallace's explanation. No, my ponderings on the songs of Curlews and Skylarks, during a quiet spell at the lek, have not resolved the dilemma. I still don't fully understand how extravagant sexual display in birds came about.

'Males having reciprocally selected the more beautiful females'

A pair of Oystercatchers now fly past the lek, screeching hysterically, as is their wont. Sky-badgers, a local gamekeeper calls them, noting not just their black and white plumage but their habit of stealing and eating Curlew eggs. Their calls are less melodious but deliriously happy to the human ear. Both sexes do these shrieking duets. Like the Black Grouse they are boldly clad in conspicuous black and white all over, with a long, thick, bright red beak and red eyes. But in Oystercatchers both sexes are identical in colour. If sexual selection is at work in this species, it seems to work on both sexes. How can that be? Mutual mate choice was a possibility that Darwin considered but did not think likely. In the *Descent of Man* he wrote:

> It is again possible that the females may have selected the more
> beautiful males, these males having reciprocally selected the
> more beautiful females; but it is doubtful whether this double
> process of selection would be likely to occur, owing to the
> greater eagerness of one sex than the other, and whether it
> would be more efficient than selection on one side alone.

In a letter written in April 1867, which he expanded into two long articles the following year, Wallace had made an acute observation: that in some bird species both males and females are brightly coloured and that these were mostly birds that breed in holes in trees or holes in the ground or in covered or domed nests where the brooding parent could not be seen by predators. He cited kingfishers, bee-eaters, rollers, woodpeckers, parrots and tits. Shelduck are unusual among ducks in that both sexes are almost identical and boldly coloured; they are also unusual among ducks in nesting in burrows. The obvious explanation is that in birds that breed in holes, the female does not need to be camouflaged while incubating so she can also dress to impress. For Wallace, this was a big flaw in the theory of female choice and strong evidence that females were under natural, not sexual selection. Yet he conceded that 'in these cases *sexual selection* had acted unchecked in both sexes', which surely supported Darwin. If there was both male and female choice going on, through mutually selective mating, with each sex picking the smartest (or least scruffy) mate it can find, unconstrained by the need for camouflage, then it vindicated the role of mate choice in general.

Yet Darwin, by the time he came to write the *Descent*, was ready to disagree with Wallace almost for the sake of it. He made a much feebler counter-argument to Wallace's point. After listing quite a few species where dull females nested in holes or bright ones did not, he then argued that kingfishers and parrots do not have brightly coloured females because they nest in holes, but they nest in holes because they are brightly coloured. 'It seems to me much more probable that in most cases, as the females were rendered more and more brilliant from partaking of the colours of the male, they were gradually led to change their instincts (supposing they had originally built open nests), and to seek protection by building domed or concealed nests.' This was scraping the logical barrel.

So the idea of both sexes in some species dressing equally smartly – and identically – during the breeding season, so as to be chosen as mates, has been out there for a long time, albeit very sketchily. Yet it was the existence of such birds in which both sexes are brightly coloured that persuaded many later biologists, especially Julian Huxley, to reject sexual selection as an explanation for bright colours except in a few highly polygamous species. How can sexual selection be at work if females are also colourful?

We now know that this conclusion was too hasty and the idea considered by both Wallace and Darwin of mutual sexual selection does indeed have empirical support. It was not until the early 1990s that mutual sexual selection was tested in an ingenious experiment in the north Pacific Ocean. Two Cambridge University scientists, Ian Jones and Fiona Hunter, were on Buldir Island, a remote Aleutian volcanic island nearer Kamchatka than Alaska. Buldir is home to millions of seabirds, including Crested Auklets. Like most seabirds male Crested Auklets are almost identical to females. Like their Puffin cousins, whose beaks grow larger and more colourful in spring, both sexes of Crested Auklet become smarter-looking in the breeding season. Small black birds with white eyes and stubby, fluorescent orange beaks, in spring they grow curved black crests from the front of their heads that bend forward above their beaks. Jones and Hunter emulated Malte Andersson by sticking extra-long crests on some models of the birds, extra-short crests on some others and then leaving these models on large flat rocks where Crested Auklets gathered to pair up at the start of the breeding season. Sure enough, females responded to the long-crested male models more than to the short-crested ones, with 'arch, hunch and touch' displays, while males likewise responded more to long-crested female models with 'ruff-sniff and touch' displays. It seemed that crest length mattered even in this pair-bonding, monogamous bird with near identical male and female plumage. And it mattered to both sexes.

Similar results came from studies of other seabirds including Least Auklets, King Penguins and more. The insight had huge implications. Darwin saw mate choice everywhere, as a significant force in many aspects of evolution, to the bemusement of his contemporaries. Those who followed him, as well as trying to dismiss and downplay the idea of mate choice altogether, steadily shrank the scope of it during the twentieth century. They said that if it did happen it was confined to female choice in polygamous species and only those in which males were especially different from females in appearance and especially brightly ornamented – though maybe not even then. Yet here was a clear case of mate choice by both sexes for an ornament grown and displayed in the breeding season by both sexes in a monomorphic, monogamous species.

Would Julian Huxley really now be so sure that the ornamentation and behaviour of both sexes of his monogamous and monomorphic Great Crested Grebes did not play a part in their getting a good mate; or, in the age of DNA fingerprinting, that they did not sometimes indulge in 'extra-pair copulations' where quality of display mattered?

King Penguins experience mutual sexual selection.

Moreover, by identifying male choice of females and female choice of males as happening at the same time, Jones and Hunter would open up a new way of thinking about human beings, those non-lekking, pair-bond-obsessed primates. As I will argue in chapter 12, human mate choice may be somewhat different from that in seabirds if males do not apply the same criteria as females. Nonetheless, that we are a species in which both sexes are choosy hardly needs saying.

'Central because they are attractive, not attractive because they are central'

It's mid June, two months after the morning at the Black Grouse lek, and I am once again writing these words while squatting on a camp chair in a small, camo-pattern hide. But this time I am in Norway, close to the Arctic Circle, and it is midnight. At this time of year this far north there is no night, just a gradual, twilight joining of dusk to dawn when the sun goes behind a hill for a couple of hours. Outside my hide, the scenery is much like at the Black Grouse lek: a boggy peat fen with patches of sedge, heather, moss and cotton grass – with flowering cloudberry and clumps of dwarf birch as exotic additions.

Despite the light, the wildlife mostly still treats these small hours as sleeping time. All except one species of bird. I am sitting right in the middle of a frenetic lek that has grown steadily more noisy as the light dimmed. At this secret site, known only to my skilful guide, Terje, the air is filled with a chorus of overlapping, wondrously strange calls. Each song starts with a sort of quiet, tuneful, twittering tinkle that sounds like a flock of distant small birds then breaks into a louder, accelerating burst of clicking like the winding of a watch or a fishing reel at the moment after a trout takes the fly. Or, and here I go again, like the soundtrack of yet another movie about aliens. The noises birds make at leks are often weird: not like typical bird songs. Not beautiful in fact.

The authors of this sound, birds called Great Snipe, are scurrying around low to the ground, from little hillock to little hillock, furtive as rodents, then sometimes flitting through the air like big moths. On reaching a preferred mound, a bird starts the tuneful tinkle without opening his beak then stands tall, stretches his neck up high, pulls up his shoulders, half-opens his wings, puffs out his breast and throws open his long beak, from which issues the clicking. Then as the clicking climaxes and stops after about two seconds, to be followed by another sort of warbling whistle, the songster suddenly spreads his white tail like a fan and turns it on its side as he stands quite still for a few moments. In the twilight, the white flash of the tail is like a light going on. The whole ritual is repeated every half minute or so.

I am surrounded by these ritually posturing, persistently calling, occasionally battling birds, some just a few oblivious feet from my own feet – which are sticking out of the open door of the hide. The clicking is continuous and overlapping. It began sporadically around three hours ago in the evening daylight and will peter out in another five hours as the sun rises higher, in the morning – when I can at last leave the hide and stumble bleary-eyed back down the mountain to breakfast and a nap. But for much of the eight hours of the non-night, these birds are hard at work lekking. It's tiring enough for me watching all night, though the time flies; for the birds it must be exhausting.

Great Snipe are small, plump, mottled-brown, nocturnal wading birds wearing streaky, dead-grass camouflage on their backs and equipped with long beaks designed for finding worms in soft mud. They returned from central Africa a few weeks ago to breed in these bogs and will go back there in August, flying – we now know – thousands of miles non-stop for three days at an astonishing average speed of nearly sixty miles an hour while sometimes reaching altitudes of more than 25,000 feet. A little larger than Common Snipe, they are the only one of seventeen snipe species that leks and one of only a handful of wading shorebirds that do.

Great Snipe warbling at the lek.

The first thing that strikes me as I watch the Great Snipe lek is how similar the whole experience is to the Black Grouse lek. Both leks are about the size of a tennis court. The location and timing of both are predictable not just from one day to the next but from one year to the next. The Great Snipe lek may move by a hundred metres or so between years but usually no more than that. Yet the terrain is unremarkable in both species: the bog where these Great Snipe are lekking looks like any other stretch of hummocky fen on these vast expanses of mountain slopes and plateaus. In both species, each male sticks to his own patch and occasionally but rather half-heartedly fights his neighbours to defend the boundaries. Each species spends a huge amount of time in ritualised display, whether or not there is a female present. Each makes two quite different and bizarre sounds, one tuneful (roo-kooing and the tinkle) and one harsher and louder (the sneeze and the clicking). Both species do flutter-jumps. Both take advantage of a lull in proceedings to have quick naps, while not deserting the territory. There is an infectious, competitive pattern to the displaying in both – one male setting off others.

There are about fifteen to twenty male Great Snipe on this lek, much the same number as there are male Black Grouse on the lek at home. As in Black Grouse, the males here can apparently be divided into dominants at the centre who display the most, subdominants around the edge and subordinates who get chased by everybody, though it's hard for me to keep track of the different individuals tonight, what with all the surreptitious rodent-running between hummocks. (The biologist Jacob Höglund and his colleagues have captured and marked males on Swedish leks to work out what is going on.) As in Black Grouse, so also in Great Snipe, it is one or two central, dominant males on each lek that achieve nearly all the matings. Males of both species frequently disrupt each other's mating attempts, or so I read. This night I do not see a female but Terje tells me that's normal. They are furtive and hard to spot among the tussocks.

Both species have close cousins living in the same region in similar habitats but which do not lek: Red and Hazel Grouse; Common and Jack Snipe. Given how utterly different Black Grouse and Great Snipe are in ancestry, appearance, size, diet and migration habits, this similarity in lekking behaviour is arresting. A Great Snipe's nuptial rituals are more like a Black Grouse's than they are like a Common Snipe's – or than a Black Grouse's are like a Red Grouse's. It seems clear that lekking, and the female choice that lies behind it, can force very different species into convergent patterns of behaviour.

But – and this is the main reason I am here on a Norwegian mountain at an hour when I should be asleep – there is one huge difference between Great Snipe and Black Grouse at the lek. In almost every lekking bird species, from Black Grouse to Sage Grouse to birds of paradise to manakins to Ruffs, the displaying males are decked out in colourful finery with arrestingly bright plumage often accessorised with bold fleshy combs, wattles or air sacs. Every bit of the bird is advertising something, like the racing suit of a Formula One driver. Not here. The male Great Snipe is perfectly camouflaged. Only the white flash in his tail is at all bright and white does not really count as a colour.

Male Great Snipe flash their white tail at the
climax of their display.

What's more, he is indistinguishable from the female. They are almost identical in size, shape and colour, though unusually, she is slightly larger than her mate. Only his white tail feathers, flashed at the end of the display, are whiter than hers.

There are two ideas to explain this lack of sexual dimorphism and colour in Great Snipe. The first is that this is mainly an audio lek, not a video lek. If what counts in the female's discrimination is the sound, not the sight, of a good male, then sexual selection might have gone to work on exaggerating the clicking and warbling rather than the plumage. Confining sexual selection to the audio channel leaves natural selection free to maintain both sexes' camouflage: Darwin gets the sound while Wallace gets the plumage, if you like. That's what Jacob Höglund thinks and he argues that since the species is nocturnal, lekking in poor light, colour would not help much. Perhaps, but at the lek I am watching it is never too dark to perceive colours, while the calls of Great Snipe, though frequent and conspicuous, are no more so than those of Black Grouse or Peafowl. That is to say, if sexual selection has been exaggerating song in this species it has not gone especially far. And males do spend a lot of time posturing, as if they were indeed dressed in gaudy plumage rather than drab camouflage.

An alternative idea occurs to me. Perhaps this species has only just begun lekking. Perhaps it is only in the past few hundreds of thousands of years or less that the species has abandoned monogamous pair bonds and started operating a system where males market themselves competitively in one space to choosy females. After all, at various points in history this must have happened to other species too. There must have been moments when a species had only just embarked on the lekking habit, however brief that moment may have been in evolutionary terms. I argued earlier that especially if Fisher is right, runaway selection for bright colours might happen very fast and not be easy to catch in the act. Are Great Snipe an opportunity to effect that capture?

In 1988 Höglund and his colleagues did an experiment to elucidate whether the white tail feathers work to attract females or repel males. They captured some birds on the lek and using dark or white typing correction fluid (remember that?) they whitened the tail feathers of some birds, painting over the black patches on the inner parts of the feathers, and darkened those on other males, filling in the gaps between the black spots. They then watched every fight and counted every copulation. There was no effect on fighting: having a whiter or a blacker tail had no impact on whether a male won fights. But there was a big effect on mating: the whiter-tailed males were chosen by more females. The females still checked out all the males but mated with the whiter ones: 'The data on female visits show that females were attracted to both experimental and control males, but when visiting the controls, the females seemed to choose not to copulate with them.' So visual cues do matter.

Earlier, Höglund did another experiment in which he removed certain male Great Snipe from the lek for one night (keeping them in a cloth bag inside his hide) and observed what happened to their territories. The birds were released the next night and seemed none the worse for this alien-abduction experience, soon resuming their territories. Perhaps surprisingly, removing the dominant birds resulted in no reallocation of their territories, which stayed vacant. Removing more junior birds caused others to move into the empty spaces. Höglund concluded that this strongly supports the hotshot model not the hotspot model of lek evolution, because the birds seemed keen to shift closer to a top male, not a top location. They put it neatly: 'Male Great Snipe are central because they are attractive not attractive because they are central.' An attractive male will draw other males to display close to him and this is how a lek develops. A female is drawn towards a central male on a lek because he displays well not because he is central.

Incidentally, in 1995 Höglund had been part of a Black Grouse study with Rauno Alatalo and Arne Lundberg that came to a similar conclusion. They found that the most popular copulation spot was not necessarily in the same location as in the previous year: it was wherever the most popular male was. In each season new territories 'were established closer to the copulation centre than to the lek centre of the previous year' and 'males close to the preferred males seemed to achieve copulations because of their proximity to such males'. Hotshot, not hotspot.

At around 2 a.m., in the Great Snipe hide, sleepiness gets the better of me and I lie down to rest for a while, my dreams filled with the insistent clicking of fishing reels. I wake two hours later as the orange rays of the sun begin to illuminate the lek, and watch a male just a few feet from me, still clicking away but with a little less urgency. How did this species start lekking and why? The ancestors of the Great Snipe probably did not lek. No other species of snipe do, so the probability is that this species descends from – or branched off from – a non-lekker. There are around seventeen other species of true snipe and yet this is the only one that displays exclusively on the ground and in groups (old reports that the Pintail Snipe that breeds in Siberia and winters in south Asia sometimes displays in the air in flocks, which might perhaps be aerial leks, have since been refuted). The closest relative of the Great Snipe is the African Snipe, which displays individually in the air like every other snipe. There is nothing in the biology of the Great Snipe to explain why it behaves so differently from all its cousins.

So here is what I think has been going on in Great Snipe. Some time in the past – maybe a few thousand years ago, maybe half a million, hard to tell – female Great Snipe returning from Africa started being attracted to males with good displays, caring nothing for whether such males had already mated. Other males then found that their best strategy was to hang around near the best males hoping for some spillover action – and being partly successful passed on this tendency to their sons. Soon the species found itself genetically predisposed to lekking. Females became more and more discriminating, perhaps because that way they stood a better chance of producing a sexually successful son. The cues they chose to discriminate were such things as frequency of display, neatness of song and whiteness of tail. Mutations that increased the number or size or whiteness of the white tail feathers thrived, and females spread these mutations around the range of the species between the Urals and Norway by returning from Africa to different leks than those where they were conceived. If so, come back in another ten thousand years and you might find the species equipped with great big, exaggerated, pure white tails. Perhaps too the middle feathers of the tail, which are currently russet with a black spot in the centre, will have become brighter red with a bolder black mark. Males will look distinctly different from females. A future reader living ten thousand years from now can check my prediction!

There are other species that look like they may be new to lekking. Buff-breasted Sandpipers in the North American Arctic lek, but in a somewhat haphazard way, with unpredictable, low-density lek sites that do not often persist from year to year, and quite a lot of solitary displaying. Males are not especially colourful or ornamented, don't fight much, look like females and unlike most lekking species share mating opportunities fairly equitably. One DNA fingerprinting study found that there were more fathers than broods in this species as a result of 'females mating with solitary males off leks, and multiple mating by females'. The authors concluded that 'sexual selection through female choice is weak in buff-breasted sandpipers'. This could be because snow cover is unpredictable in their preferred habitat so from year to year they have to be flexible in where they settle, but I would suggest that it might also be that they have only just started lekking – by which I mean maybe a few thousand years ago.

Perhaps the same applies to an enigmatic species of lekking grouse in North America, the Sharp-tailed Grouse. I have never managed to see a Sharp-tailed Grouse lek, though I have watched the species and visited a lekking site, right in the middle of the Little Bighorn battle-

Greater Prairie Chickens strut, dance and fight on the lek.
Photo by Matthew Dryden.

field close to the spot where General Custer died, but I was just too late in the season for the lek. The species is unusual in several respects. Compared with its close relatives, the two species of so-called Prairie Chickens (though they are actually grouse), the Sharptail species seems to lek in a half-hearted way. Greater Prairie Chickens gather at regular leks to strut, dance and fight, much like Black Grouse, with inflated purple skin sacs on the side of the neck and vertical feathers held erect behind the head. For Sharptails, however, leks are unstable and unpredictable; males often display alone, are not much bigger than females and the plumage of the two sexes differs little. The male has a vigorous display in which he tap-dances in place with his head down, clicking loudly like a fishing reel and stamping up and down rapidly, with tail erect, wings spread, yellow eye combs showing and purple neck pouch swollen (I know this because there are plenty of videos online). But once he folds his ornaments away he looks just like a female, well camouflaged against the dry vegetation of the prairie-forest edge where the Sharptails live. In one study of Sharp-tailed Grouse, 'dominance status could not be predicted on the basis of body size, comb size, or age class' and 'on average, smaller males apparently obtained more matings'. I think this sounds like a species that is either fairly new to lekking, or perhaps the opposite: in the process of giving it up. As I keep saying, there is every reason to expect that some species are newer to lekking than others. All species are still evolving and the current design of ornaments on a Black Grouse or a Peacock is not the last word.

8.

Handicaps and Parasites

Well, honest John, how fare you now at home?
The spring is come, and birds are building nests;
The old cock-robin to the sty is come,
With olive feathers and its ruddy breast;
And the old cock, with wattles and red comb,
Struts with the hens, and seems to like some best

JOHN CLARE, on John Clare

8 a.m., April, the Pennine hills

In the 1980s, persuaded at last that brightly coloured male birds had been made that way by generations of choosy females, and impressed by Malte Andersson's widowbird experiment into believing that this process could not just maintain but drastically exaggerate features of plumage, zoologists began to think hard about how to measure the differences in health and strength between males that females must be choosing. A Peacock's train is, let us assume, a measuring device for announcing the relative value of the male as a mate. But what is being measured? Here the old lek paradox reared its head again. Birds that lek, like Black Grouse, Peacocks or birds of paradise, are among the most gaudy and decorated species of all. They are also the most choosy, with just one male getting the bulk of matings at any one site. But why bother being so choosy in such species? In a monogamous Red Grouse, where the pair bond may last for six months, the female should be all the more careful to choose a good, healthy male since she is going to be relying on him to defend a patch of nutritious heather and then hang around and help rear the chicks. In the Black Grouse, with its two-second pair bond, his health is going to make little or no difference to the rearing of the brood. The only difference it can make is to the health and vigour of the offspring themselves through their genes.

So in lekking species the female has ostensibly less reason to be choosy, not more. If all you are getting from the male is a dollop of sperm, you don't actually need him to be big, strong, diligent or faithful. Worse still, if it's good genes to pass on to your offspring that matter, it's a bit odd to be so obsessed with being choosy in a species that keeps minimising genetic variety by putting the population through a narrow genetic bottleneck in every generation, reducing the variation in quality between males. As Gerald Borgia pointed out back in 1979 when he coined the phrase 'lek paradox', the choosiest species have the least reason to be choosy.

One way to resolve this paradox would be to find some feature of the bird's life that keeps changing from generation to generation. If there is a fresh reason to choose slightly differently in every generation then what is best in one is less good in the next. In the clothing industry, fashion changes all the time. What is in vogue one year is old hat the next. Yet, clearly this is not true of the ornament itself in birds. Females are not suddenly deciding they prefer short tails this year. The consistency with which females have preferred the same or similar adornments and dances for decade after decade is proved by the fact that I can watch a lek and find my notes interchangeable with those of Edmund Selous.

With the work of Lande, Kirkpatrick and Andersson, female choice, sexy sons and the runaway hypothesis were back on the table in the

1980s when I started working on the topic at Oxford University. I was a convinced Fisherian. The very arbitrary randomness of bird ornaments seemed to me to demand an arbitrary, random explanation of the kind Fisher had provided. If it was a matter of selecting the fittest males there would surely be one or a few ways of birds advertising that, and the features would be prosaic information devices, rather than baroque adverts.

But I was surprised to find many of my colleagues unpersuaded. They generally seemed to think it more likely that there was method in the madness of female birds, that the choices they made were sensible in terms of the quality of genes they got for their chicks. Picking the healthiest strongest male, who had won the most fights, made perfect sense for females that wanted only sperm. If, instead of marrying a man, a woman goes to a sperm donor, she's likely to want to know that he is at least reasonably healthy, good-looking, intelligent and athletic. So why not a bird? But I sometimes wondered if this preference for good-genes explanations was not just a general frustration that I have often noticed in science at any explanation of how something works in the world that is arbitrary and meaningless. People do long for things to make sense. And to give opportunities for experiments. Another problem with Fisher is that he leaves you less work to do: problem solved, he says, and since it's random it will have a different look in every species.

'Sexual displays endanger their performers'

Perhaps the most extreme of the suggestions as to why ornaments on males made 'good sense' after all came from an Israeli scientist named Amotz Zahavi. In 1975 he argued that the cost and risk of the ornament were the whole point. If the trait is costly and risky, the male has been put to the test. The fact that a Peacock has survived despite having 'excessive tail plumes', as Zahavi put it, proves to the Peahen that he is a knight in shining armour who has run a gauntlet of dangers and survived, so he would make a great father. 'Many, if not all, sexual displays endanger their performers. Many of them seem to be designed specifically for that purpose,' Zahavi wrote. Rejecting Fisher's logic out of hand, he asserted that sexual selection is 'effective only by selecting a character that lowers the survival of the individual'. In a sense, he brought sexual selection back under the command of natural selection again, as Wallace had done a century before, by making successful reproduction a means to ensure the superior survival of offspring.

Zahavi's idea closely resembles, and is logically similar to, Thorstein Veblen's 1899 theory of 'conspicuous consumption' by rich human beings. Extravagant spending on jewellery, art and fashion by the nouveau riche was wasteful, Veblen argued, but that was its very point. To throw away a fortune on a useless object is to show the world how rich you are. That this has a directly sexual result in human beings, in the form of mate choice opportunities, did not form part of Veblen's argument but it is pretty obvious to most observers.

Zahavi's argument as an explanation of male ornaments in birds was supported by neither experimental evidence nor mathematical proof and it soon ran into a hail of criticism. It struggles to explain the great diversity of male ornaments and displays, let alone why an Argus Pheasant should have optical illusions of spheres on its wings – these are presumably no more costly or tricky to grow than many other patterns. It leads to the logical conclusion that males should be born with one eye or one leg, the better to handicap them, as Richard Dawkins pointed out. Richard Prum recalled a 1970s advertising slogan for the American jam manufacturer, Smuckers. 'With a name like Smuckers, it has to be good!' Why not then, a team of television comics asked at the time, call it death camp jam, dog vomit preserve or painful rectal itch jelly? Then it would have to be really, really good to sell. The logic is faulty.

Besides, as Dustin Penn of the Konrad Lorenz Institute in Vienna later pointed out, Zahavi had been forced to contradict himself because 'on the one hand he argued that signals have extra costs that make them wasteful, but on the other hand he maintained that these additional costs functioned to demonstrate their honesty'. If so, and these costs provide lucrative rewards in the mating market, they were not being wasteful after all. Penn concluded that Zahavi's argument was essentially circular because he began by saying that costly ornaments are reliable signals of male quality, before concluding that reliable signals evolved because they were costly. Then, as Helena Cronin put it, an animal that faces a trade-off between mating and risking being eaten by a predator is no different in principle from an animal that faces a trade-off between foraging for food and risking being eaten by a predator. For reasons like this, throughout the 1980s Zahavi's idea remained a bit of a laughing stock within evolutionary biology. 'In polygynous species the process envisaged by Fisher is overwhelmingly more important than any kind of handicap effect,' wrote the veteran evolutionist John Maynard Smith in 1985. In short, a successful male benefits from growing an ornament to attract a female despite it being a handicap to survival, not because it is a handicap.

'Rhyme and reason'

Why then am I beating this dead horse? Because in 1990 suddenly and enduringly a form of the Zahavi idea – or at least what claimed to be a form of it – sprang back to life and gathered new support not just in the study of sexual selection but throughout biology and the social sciences, in the guise of honest signalling theory. The cause was an influential paper by a mathematically skilled Oxford evolutionary biologist (and former colleague of mine) by the name of Alan Grafen. His paper took a secondary suggestion that Zahavi had made in 1977 and ran it through a mathematical model. This was the idea that males might be adjusting the performance of their sexual displays according to their fitness. A weedy male might struggle to display large ornaments and thrive whereas a strong one might grow bigger ones and give them a more convincing shake. Secondary sexual ornaments such as Peacock trains might be condition-dependent so that discerning females could read the quality of the male from the colour, shine, intricacy, size or movement of the male ornament. The fact that such ornaments are difficult to grow or to display is then a key part of their appeal to evolution and to females. As Grafen put it, 'according to the handicap principle' there is indeed some 'rhyme and reason' in sexual selection 'in contrast to the Fisher process, in which the form of the signal is more or less arbitrary'. The Fisher hypothesis was, said Grafen, 'too clever by half'.

A generation of evolutionary biologists, influenced mostly by Alan Grafen's model, came to think that the handicap principle had now been proven right. In Grafen's words, 'Zahavi's major claims for the handicap principle are thus vindicated.' John Maynard Smith, Richard Dawkins and Bill Hamilton, all leading evolutionary theorists, fulsomely endorsed Grafen's apparent vindication of Zahavi. Many others, myself included (at the time I was writing a book called *The Red Queen* that covered some of this ground), followed their lead. I was, as I later put it in an email to Richard Prum, 'grafenised'.

Then in 2015, Dustin Penn entered the fray. He was working on sexual selection and communication (pheromones and ultrasonic vocalisations) in wild house mice when he came across a couple of papers by Szabolcs Számadó of Eötvös University in Budapest that challenged the idea that Grafen's model was applicable to the real world. Penn invited Számadó to Vienna to debate the matter and they soon concluded that Grafen's model was not a handicap model at all, but an idea known as the 'condition-dependent' hypothesis. As Penn told me, 'Because this idea was proposed by Zahavi, everyone just assumed that it was a handicap hypothesis. We realised that it is completely different and we explain why. Alan had accepted this widespread assumption. Thus, we realised that there is no theoretical or empirical support for the handicap principle, and that it can be rejected for many reasons.'

They argued that Grafen's model did not support Zahavi's claims that it was costliness of ornaments that made them honest. Indeed, the whole point is that the really fit male does not find the ornament too costly. Grafen was not arguing that the cost is the purpose of the ornament, the thing that makes it revealing. The males, far from maximising the handicap to show their quality (and presumably watch the handicap kill off their weaker rivals), are minimising the trade-offs involved in display so as to survive despite displaying. It's really just the old notion going back to both Wallace and Darwin that females might be more willing to be easily seduced by the fittest males, which ensures they buy the best genes. 'We do not consider this idea to be a type of handicap hypothesis,' commented Penn and Számadó.

Not surprisingly, Grafen disagreed with much of Penn's and Számadó's criticism. They invited him to speak at a symposium in Estorial in Portugal in 2017 – the year that Zahavi died. Penn says they could not convince him that there was a problem. 'He told me that we were "tilting at windmills", and his final words at the end of the talks were something like, "I still stand with Amotz."' In 2018 Grafen published a new study based on models that made the argument that sexually selected traits got exaggerated when they correlated only weakly with male quality: if they correlated strongly there was no need to exaggerate them. Yet surely we have a better explanation for when traits get wildly exaggerated: it happens in highly polygynous species like Black Grouse, where males are excused all parental duties and female choice is very strong.

After writing a couple of critiques together and organising symposia at conferences, Számadó and Penn wrote a lengthy review to explain the history of the handicap principle entitled 'The Handicap Principle: How an Erroneous Hypothesis Became a Scientific Principle', which was published in the journal *Biological Reviews* in 2020.

Penn and Számadó conceded that Grafen had certainly showed that under certain assumptions, a condition-dependent ornament could be useful to a female in choosing a good male. But this was not really new and it could not in itself rule out the Fisher runaway process taking over and causing exaggeration of an ornament that did not necessarily tell the truth about the quality of the male. Grafen did not, as he claimed, prove the value of the ornament being wasteful and extravagant under realistic assumptions. For a start, and to repeat the point, if the 'wasteful' extravagance of the ornament is what has earned the male more matings, then by definition it wasn't wasteful. Also, mathematical analyses showed that a signal being costly is neither a necessary nor a sufficient guarantee that it will be honest, a key assumption of Grafen's version of Zahavi. He was not able to show that less costly or less honest ornaments could not also be successfully used in mate selection. Nor, finally, had it been demonstrated whether the cost to a female of bearing sons with disadvantageous ornaments was outweighed by the reward of seeing them get more matings – a metric on which the Fisher theorem was easily validated.

'It takes all the running you can do to keep in the same place'

Handicaps are on the agenda as I watch Wonky Tail display at the lek. This is (I repeat) his third year here and he is in his prime. He carries his entire tail, indeed his entire rear half, at a steep angle so that the left side of the longest tail feathers trail in the frost and look a little bedraggled. He has at last had a single mating this year, as recounted in chapter 3, the first I have seen, despite still being not quite at the centre of the lek. The fact that his display is far from symmetric thanks to his disability does not seem to detract from his enthusiasm and energy. But perhaps he would have achieved more if his tail was less wonky. Or perhaps his very handicap helped him get at least one mating from a sympathetic female who looked him over and said to herself: 'By Jove that fellow's brave: displaying with vigour even though he's got a significant disability; I think he deserves a roll in the hay.'

The significance of this theoretical dispute is that there was a time when Alan Grafen and Amotz Zahavi seemed to have removed the Fisher runaway process from the running. Once again, like Wallace and Huxley, they had killed the idea. Despite the lack of abundant proof of his own Zahavi-Grafen process, Grafen went as far as to claim that 'to believe in the Fisher-Lande process as an explanation of sexual

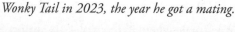

Wonky Tail in 2023, the year he got a mating.

selection without abundant proof is methodologically wicked'. As Prum put it to the philosopher David Rothenberg, biologists seemed to 'prefer to obfuscate the whole thing by arguing that females are constrained by a need to select for fitness, instead of according to the whims of fashion'. Biologists, argued Rothenberg, are afraid of true sexual selection because it leads to arbitrariness. Thus in the 1990s Wallace eclipsed Darwin for a third time.

The effect of Grafen's model was to inspire a generation of biologists to fan out across the world to measure the cost of growing tails or singing songs. Their hypothesis was in effect that females were being thoroughly eugenic in their choice of the best male. In particular they began to suspect that ornaments left males with weaker immune systems more vulnerable to parasites, so the ornaments sorted the wheat from the chaff. 'Good genes' in the immune system – that was the reward that female Black Grouse got for picking the gaudiest male.

But, but, but … what about the lek paradox? What's the point of choosing a healthier male when all the males are almost equally healthy? Females picking the best males will soon run out of genetic variation. Or will they? Perhaps, hidden beneath the surface, something changes so that the most beautiful male is one with a different set of genes each time. There are two things that could explain such a cyclical advantage. First, mutations: every individual carries brand-new minor mutations in its genes, most of which are harmful. Perhaps this is enough to give females something to work at. Second, parasites. Bill Hamilton, the evolutionary biologist working in Oxford whose novel genetic insights had equalled those of Robert Trivers in the 1960s and 1970s, argued that diseases caused by bacteria, viruses or other parasites are in a continual arms race with their hosts. One side constantly improves its defence, the other constantly hones its attack. We saw this clearly during the pandemic: one strain of the virus displaced another within a few weeks to evade the immunity acquired by much of the population, and the one that appeared in late 2021, omicron, was so infectious that the following year even the Chinese government had to admit defeat and give up trying to use lockdowns as a defence.

But the molecular defences change too. As one strain of a virus provokes antibodies in its victims, so the evolutionary pressure on the virus to mutate into a new strain that evades those antibodies intensifies. The new strain kills or debilitates the most susceptible hosts, leaving ones that happen to have good defences against the new strain to populate the next generation. Something like this has been happening with avian influenza in seabirds as I write: a lethal new outbreak in 2022 left many dead; the survivors include some lucky birds that missed the infection, but also some that had genes that rendered them more resistant. And sure enough the next summer, 2023, seabird mortality was lower and these resistance genes presumably commoner. To put it simply, the host keeps changing the locks, the parasites keep changing the keys.

So ruthless is the competition among strains of a parasite that a disease effectively has to choose one key. Covid switched from alpha to delta then omicron almost completely – it did not 'keep its options open'. But the advantage is temporary. A strain of disease that becomes dominant sows the seeds of its own demise by giving selective advantage to whichever strain of host best resists it. So the triumph is necessarily short-lived. A host might have an old lock that was vulnerable in a previous generation but is now good again.

The race between animals and their parasites is a race in which – as the Red Queen puts it in Lewis Carroll's *Alice Through the Looking Glass* – 'it takes all the running you can do to keep in the same place'. Carroll wrote his surreal fantasy, the second Alice book, in Oxford in 1871, the same year that Darwin published the *Descent of Man*. Neither man can have imagined that the two might come to cross-fertilise each other more than a century later. In the 1980s, as I chronicled in *The Red Queen*, an evolutionary biologist named Graham Bell borrowed the scarlet sovereign's remark as a metaphor for the tendency of parasites and their hosts to engage in perpetual arms races in which neither side gained more than a temporary advantage but both sides were continually changing their genes.

In such a system, there may well be a good reason to pick a vigorous and colourful male in the hope that he possesses the latest resistance genes against the latest version of the prevailing parasite, or even an old resistance gene that has come back from the dead and is effective again – and in the hope that these resistance genes will still work in the next generation at least. Much depends on how fast the changes will be. If the cycle is very short, so that the offspring of resistant parents are already much more susceptible, then the female should pick a sickly male in the hope that his genes will work much better in the next generation. But if it is longer, then healthy males will generally have healthy sons. Bill Hamilton and Marlene Zuk wrote a paper in 1982 setting out this argument and suggesting that it could resolve the lek paradox, give a reason that never runs out for picking the best mate and explain the evolution of colourful plumage.

To test their idea, Hamilton and Zuk predicted that whereas within a species the most brightly coloured males should have the fewest parasites, between species the most brightly coloured species should be the ones most plagued by parasites. Their argument was that species would evolve bright colours in order to try to enable females to select less infected males in those species where such infections are a particular problem. They consulted seven surveys of blood parasite incidence in 109 North American bird species and found a weak but significant correlation between colourfulness and parasite load. Brightly coloured bird species tend to be the ones that are more troubled by protozoa and nematodes in their blood. Strikingly, birds that live on islands tend to be much duller than the continental species from which they are descended, as if colonising a remote island removes the need to be flashy, and of course such island pioneers do tend to leave at least some of their parasites behind. Hamilton and Zuk suggested that birds are choosing their mates using the criterion of genetic disease resistance by scrutinising features of plumage 'whose full expression is dependent on health and vigor'.

A new wrinkle was added to the theory in 1992 when other scientists suggested that testosterone has the effect of suppressing immune function. So the more testosterone a bird pumps out of its glands to enlarge its feathers and brighten its colours, the more vulnerable it makes itself to parasites. This makes the plumage an 'honest signal'. You cannot fake it till you make it on the lek. If you try to big up your colour too much you will fall sick. It is certainly true that diverting energy into growing large feathers and behaving aggressively must have an opportunity cost and be harder for sickly males to do but if that is the purpose of bright plumage then why is it so infinitely variable between species? Surely one or two reliable colours, and one or two reliable features, would prove to be the most effective at demonstrating health and would come to dominate the world of extravagant display plumage?

'If belief in the alternative hypothesis is "wicked", there's little choice to make'

These thoughts were running through my mind when in late March I found myself sitting, shivering on a bench in an old shipping container by a roadside in western Colorado. It is minus 9 degrees, dark and the snow lies deep, crisp and even. Despite the chill, one side of the container has been opened so we can see across the fields to where a lek of Sage Grouse is about to happen. Sage Grouse, a little bigger than Black Grouse, are in some ways the greatest lekkers of the lot: they will gather in groups of one hundred or more and sure enough this morning as the light comes in we spot a flock of eighty birds some distance away, doing their extraordinary thing. Each male fans his spiky tail

over his back, swells his neck into a vast and pendulous white ruff and
struts about, periodically bouncing two bare-skinned olive-green
patches through the feathers of the ruff with a pneumatic noise of
somewhat comic quality.

This was not my first experience of a Sage Grouse lek. In 1988 I also
shivered in the chill of a sharp wind in an April dawn on the Wyoming
prairie, as my old friend the evolutionary biologist Mark Boyce – a
subtle-minded theoretician with the outward appearance of a
back-country fur-trapper – explained the manoeuvres of these extraordinary birds strutting their stuff just a hundred metres away. Three
years before, another friend, the cerebral, shy evolutionary theorist Bill
Hamilton himself had visited Boyce to watch a Sage Grouse lek. The
two biologists discussed whether the Red Queen might be at work on
a Sage Grouse lek, giving the advantage to a different set of genes every
few years and requiring females to be highly selective in the hope of
choosing the most disease-resistant genes for their offspring. Boyce
became one of the first scientists to test Hamilton and Zuk's theory: he
established a captive population of Sage Grouse and set them up in
artificial leks, as described in chapter 3: a six-sided cage with one female
in it and a male in a cage on each of the six sides.

In his captive leks, Boyce treated half the male Sage Grouse with a
standard antibiotic and recorded that this treatment resulted in more
frequent strutting displays and a greater chance of being preferred by
the females. He then tested Sage Grouse in the wild for avian malaria,
a protozoan parasite carried by mosquitoes, and found that infected
males attended leks less frequently, achieved fewer copulations than
expected and mated later in the season if at all. Boyce was later
intrigued to read that avian malaria tends to cause sickness early in
the morning, making any ill males more likely to show symptoms
around the time of lekking. He and his colleagues also observed fewer
lice on birds that succeeded in mating. When they painted fake
haematomas – red marks of the kind left by louse bites – on the air

sacs of captive male Sage Grouse the effect was to diminish their success at mating. All in all, then, there does seem to be a clear preference in Sage Grouse females for males with fewer parasites, however they detect it. But is that the ultimate reason for their preference, or a mere correlation?

There is a puzzle here. Hiking through a forest in Norway recently, I was lucky to glimpse a plump, dull-coloured bird scuttling off through the bilberry plants: a Hazel Grouse. Not much smaller than a Black Grouse, more wedded to dense forests, perhaps preferring hazel to birch, but otherwise just as dependent on buds and twigs to get it through the winter, the Hazel Grouse is a cousin and neighbour of the Black Grouse throughout Eurasia. Yet males and females look very similar, form pairs and do not lek. Likewise in the Pennines, as I keep emphasising, Black and Red Grouse overlap in the same habitat, but Red Grouse are monogamous, monomorphic and non-lekking.

Does it really sound convincing to argue that Black and Sage Grouse are plagued by malaria or some other blood borne disease, but Hazel and Red Grouse are not? That the reason the Hazel Grouse and the Red Grouse do not lek is because they do not have to worry about parasites and enabling females to detect disease resistance? Put like that, the theory is not persuasive – yet that is essentially what Hamilton and Zuk were arguing when they predicted that brightly coloured birds would prove to harbour more parasites. The Black Grouse, living at lower densities and ranging further afield at different seasons, is probably less likely to be parasite-plagued than the sedentary Red Grouse, which constantly defecates on its food plant, or the Hazel Grouse, which mostly stays put in the same home range.

In the grouse, at least, there's a better way of predicting which species will lek. The lekkers all live in open habitat: sagebrush for Sage Grouse, prairie for Greater and Lesser Prairie Chicken and Sharp-tailed Grouse, and moorland or forest clearings for Black Grouse. But then there's the three ptarmigans, of which Red Grouse is one, which inhabit open tundra or its alpine equivalent and don't lek. And there's manakins and birds of paradise, lekking in forests.

All right, let's steel-man the argument rather than straw-man it. If you instead argue that Red Grouse and Hazel Grouse females have a much more urgent and salient reason to pick the best available single males, parasites be damned, then I might be convinced. That reason is that by inspecting males' territories and choosing the ones with the best food, a female Red or Hazel Grouse gets to monopolise a decent food supply, which matters far more. In Black Grouse, with no such prize on offer, it's decent genes alone she seeks and those red combs, white bums, curved tails and blue neck sheen are ideally suited to giving away the disease-resistant status of the male, like that beastly little line on a Covid test. So goes the argument at least. The very fact that a male Black Grouse feels the need to show off every part of his plumage could be relevant here, because the more bright features the male has, the more likely it is that an infection would show up some-where on his body to spoil the brightness. A perfect white bum may get soiled in a bird with the runs, for instance, or the blue sheen on the neck might tarnish on a bird with lice. It's a theory at least.

In 1999 Anders Møller, a Danish biologist based in Paris at the time, together with two colleagues, published a comprehensive review of the parasite theory of sexual selection. They drew together all the studies they could find and teased out their conclusions. They found that many studies detected little or no effect, but instead of seeing this as refutation of the hypothesis, they concluded that it might be because scientists had tested the wrong parasite – after all, birds are afflicted by many different pests and it is not easy to tell which is the most impor-tant to a choosy mate. Where there was a connection between parasite load or immune function and bright colours, it did tend to support the idea that colourful species have more of a parasite problem, but that colourful individuals have the problem better under control. For instance, by comparing pairs of similar bird species, one of which has

colourful males and dull females while in the other the sexes are similar, Møller was able to show that the colourful species had a larger
spleen and other immune organs, implying a stronger immune system:
'Dichromatic species generally had larger immune defences than
monochromatic ones.' But it was not a strong effect. Møller also raised
the possibility that female choice for bright colours, by avoiding parasites, might put pressure on parasites to become more virulent.

But here is my problem with Møller's papers. They do not mention
Ronald Fisher. Not quite true, because Fisher is cited once in the different context of a method of statistical analysis, but Fisher's runaway
sexy-son hypothesis is not mentioned at all. Since it is the alternative
hypothesis, and the one that is being allegedly rejected in favour of a
good-genes theory, it was odd not to at least bring it up. I am not
picking on Møller. There has been, as Richard Prum has pointed out,
a tendency among the followers of the various good-genes theories to
ignore the runaway idea altogether, assume that female choice is always
and everywhere a matter of picking good genes for the health of her
offspring using plumage handicaps – and conclude that the task of the
scientist is merely to find which genes and which parasites. In this they
often take their cue from Alan Grafen. Here's how Prum put it: 'Even
though Grafen merely demonstrated that there were conditions under
which the handicap principle could work, he so discredited the
Fisherian theory that most evolutionary biologists concluded that the
handicap principle not only could work but would work – all the time.
If belief in the alternative hypothesis is "wicked", there's little choice to
make. Adaptive mate choice has dominated the scientific discourse ever
since.'

'Freedom begets beauty in nature'

'The reason birds are so beautiful,' says Richard Prum, munching on a tuna melt sandwich, one cold January day in New Haven, Connecticut, 'is because they don't have penises.' Prum is the William Robertson Coe Professor of Ornithology at Yale University and Head Curator of Vertebrate Zoology at the Yale Peabody Museum of Natural History. I stop, a french fry halfway to my mouth. 'Run the logic by me again.' He explains that ancestral birds almost certainly once had penises as mammals do. We know this because certain groups of birds still do – tinamous, ostriches, kiwis, some ducks – and these branched off from other birds early in the history of birds. Also, early in the development of male chickens, penises appear then shrink away. So both in ancestry and in development, penises come first. Most birds as adults have no penis, just a simple unisex cloaca from which sperm is extruded into the similar opening on the female. This has the effect of making it much harder for male birds to forcibly copulate with females. In some ducks, which do have penises, forced copulation or rape is still fairly common and it's a brutal process that can even be fatal. Why did penises disappear in most birds? Because of female choice, Prum argues. Females selected males with shorter penises till they were altogether gone, as a means – unconsciously – of gaining control over mating. 'What did female birds do with this freedom of choice?' asks Prum, taking another mouthful of tuna. 'They chose beauty. Freedom begets beauty in nature and that's a completely scientific statement.' The effect of runaway female choice has been to turn male birds, in many species, into well-mannered dandies.

It was my old friend Tim Birkhead, a bird professor and bestselling author, who helped to unravel the strange story of the duck penis. By dissecting different species of bird, he had noticed that 'extreme genitalia are the hallmark of extreme sperm competition'. That is, in birds where multiple males routinely mate with one female, setting up a sperm race inside the female, there is often complicated apparatus in one or both sexes, designed presumably to affect the outcome of the race. In 2001 a photograph was published in the journal *Nature* of a dead specimen of a rare duck, the Argentinian Lake Duck, from which extruded a truly gargantuan penis, forty-two centimetres long when stretched out. Kevin McCracken and his colleagues had been dissecting Argentinian Lake Ducks and finding large, coiled penises but this was the first time they had managed to catch one in an 'aroused' state, with its hydraulic, lymph-driven structure everted like a long sock. It was twice as long as they expected and indeed twice as long as the bird. Birkhead was intrigued, so he recruited the biologist Patricia Brennan to Sheffield University to study duck copulation in detail. She began by carefully dissecting a Mallard's vagina and found it too was highly peculiar, with branching, blind-ended pouches and a tight spiral near the end. Birkhead and Brennan called an expert on duck reproductive anatomy in France (as you do) and told him what they had found, asking him if it was typical. He was surprised, having never examined a Mallard vagina in detail, but called back a short while later, after dissecting a duck, to confirm that they were right. It's just that hidden inside a mass of connective tissue, the Mallard vagina had remained unexplored, like the surface of Venus, till the twenty-first century.

Brennan next dissected sixteen species of duck and as expected found that 'in species where males had a large phallus, the female had an anatomically complex vagina'. She and Birkhead had stumbled on an evolutionary arms race. They wrote: 'Since females cannot always escape from the sexual advances of extra-pair males, they can regain some control over fertilisation by counteracting the male's phallus, and appear to have done this by evolving a longer or more structurally complex vagina. Males, of course, respond to a more complex vagina by producing an even longer phallus.' The blind pouches serve to trap the penis into dead ends; and 'the vaginal spirals are twisted in the opposite direction to the spiralled phallus, so it would be hard to imagine a more effective way to avoid a forced insemination'.

Patricia Brennan then teamed up with Richard Prum at Yale to study duck copulation in action. She found a duck farm in California where they were crossing Pekin ducks (a breed of Mallard) with Muscovy ducks (a different species), using artificial insemination. The farmers allowed the drakes to mount females but placed glass bottles over their cloacas just before they everted their penises, and collected the sperm. Using high-speed film Brennan recorded how the Muscovy drakes explode their enormously long, corkscrew-shaped, hydraulic penises into the female vaginas at three miles an hour then lodge them there for a third of a second till ejaculation is over, using ribs, ridges and teeth 'like the pitons a mountain climber uses to maintain progress up a forbidding cliff', as Prum puts it, before deflating and withdrawing. (I am quite sure that is the weirdest sentence I have ever written.) But Muscovy ducks, like Mallard, have thick, twisted, corkscrew vaginas full of blind cul-de-sacs, which spiral clockwise, in the opposite direction to the penises to ensure that unwanted sperm from a rapist male almost always fails to reach its intended destination (the second weirdest sentence). It seems like we have caught the ducks in the middle of an evolutionary war between the sexes to decide who gets to control the fertilisation of eggs. In almost all other birds, the loss of the penis has firmly placed females in charge of one aspect of evolution. Having wrested the power of conception away from males, female birds have found themselves free to breed exactly the kinds of males that suit them, and they have chosen beauty in many cases. So goes Prum's logic.

I am immediately struck by a parallel between the duck penis and the train of the Peacock, or the lyre of the Black Grouse or the dance of a manakin. The arms race in these cases is psychological rather than physical but the effect is the same: exaggeration in one sex to overcome resistance in the other. The more discerning females become, the more persuasive males evolve to be. The enormous penis of the Argentinian

Lake Duck is surely a product of exactly the same kind of runaway sexual selection as the train of a Peacock.

The reason I am here to talk to Richard Prum, however, is not mainly to discuss duck sex, but to understand the role he has played in passionately, perhaps even aggressively, attempting to wrench the sexual selection debate back to the direction Darwin was taking it in a century and a half ago. Prum thinks the cause of sexual selection is that 'beauty happens'. We have been far too utilitarian about it in recent years, far too in thrall to Wallace and Huxley and Zahavi and Grafen and not nearly as interested in Fisher's and Darwin's runaway, random exaggeration of seduction for its own sake as we should be, he argues. And we forgot about beauty. In 2017 Prum published a book, *The Evolution of Beauty*, which set out the results of his own research mainly on lekking manakins in South American rainforests and argued that biologists were getting sexual selection wrong.

That January day in the Peabody Museum at Yale University before lunch, Prum beckoned me into a vast hangar with a seemingly endless line of huge rolling cabinets, each filled with drawers in which lay the beautifully preserved skins of more than a hundred thousand dead birds, all carefully labelled. Opening a drawer, he picked up an absurdly fluorescent orange bird with black wings, an Amazonian Cock-of-the-Rock, famous for its forest leks. 'We've discovered a bizarre thing about this bird,' he said. 'The carotenoid pigment it uses to colour its feathers is a slightly different chemical from the one used by its closest relative from further north' – he gestured at a slightly paler but otherwise similar bird – 'which is different again from the one in the feathers of their cousin'. He pointed at a smaller, reddish bird. The chemicals are all carotenoids, fat-soluble pigments used to turn feathers or skin red or yellow. If synthesising enough carotenoid is a test and a handicap, why would different versions appear in different but closely related species? This diversification of ornaments for its own sake is a hallmark of sexual selection.

'It had actually become impossible to be a contemporary Darwinian'

Let me step back a short way. Russell Lande's mathematical model of Ronald Fisher's sexy-son hypothesis in 1980 restored the view that Fisher might have been right after Huxley's subtle denigration. Ornaments were about making mates attractive because attractive mates left more descendants, and that is all there is to it: no need for ornaments to be badges of health. There followed a minor gold rush of mathematical models to test the limits of the idea. One of the results of this, which went largely unnoticed by the experimentalists in the field, was to prove that sexy-son and good-genes theories were not opposites at all. As Louise Mead and Stevan Arnold put it in 2004: 'It is misleading to paint the basic Fisher process and good genes as alternative explanations for ornament exaggeration, as is common in many textbook accounts. Rather, the issue is whether good genes do their work alongside the inevitable Fisher process.' Mead and Arnold were arguing that the runaway version of sexual selection was so powerful, so inevitable once female choice got started, so likely to proceed with 'ever-increasing speed' in Fisher's words, as to render the debate about good genes versus sexy sons moot. Even if females are getting a direct benefit in terms of disease-resistant genes by selecting a brilliantly coloured male, they can also be getting an indirect benefit in terms of attractive sons. And the latter will swamp the former. Arnold again: 'Even when the genetic correlations that are necessary for the good genes process are present, the supplementary exaggeration of the ornament as a result of good genes might be relatively small.' They added that the ubiquity of the runaway process, with its correlated evolution of ornament and preference, 'will surprise many students of sexual selection who have incorrectly viewed the runaway as an alternative to good genes or sexual conflict models'. Here is how I have come to think of the issue: the good-genes theory says that the females are choosing flamboyant males so that their young are healthy; the Fisher theory says that they are choosing flamboyant males so that their young are sexually attractive. Both can be true at the same time.

The models suggested that Fisher's sexy-son effect can explain the multifarious exaggeration of ornaments and feathers that seems to happen in so many species. 'The models indicate that, even when there is direct selection on preferences, the basic Fisherian process can easily lead to the evolution of multiple ornaments and preferences,' as Mead and Arnold put it.

Building on this and other insights, Prum began to argue that we had been placing the burden of proof the wrong way round. Far from needing to demonstrate the runaway, sexy-son effect in the field, it should have been the default, the null hypothesis. Fisher's theory should be the explanation of first resort, not last.

Prum is the latest biologist to take up the cudgels on behalf of Darwin against the forces of Wallace, in the tradition of Selous, Fisher and Lande. As the above passage illustrates, he is also the most radical. But unlike the mathematical modellers such as Fisher, Lande and Arnold, Prum is a field biologist. For many years he has studied small fruit-eating birds in South American forests called manakins. These brightly coloured finch-sized birds have elaborate and bizarre leks, performed on or above branches or stumps in clearings in the forest. Some involve two or more males performing choreographed joint dances, while other species have developed such weird noise-making wing feathers, which produce rapid, loud, clicking sounds, that their very wing bones have changed shape into thick and warped bows – inconvenient for flight but vital for seduction.

Prum thinks Darwin was absolutely right to anthropomorphise birds as having a 'taste for beauty', that Fisher was dead right about how display evolves through runaway selection and that there is a direct link from bird plumage to human aesthetic judgement about beauty, too. His book *The Evolution of Beauty* infuriated some colleagues, but thrilled others.

In 1997 Prum sent a paper to a high-profile journal describing the dances performed by various manakin species and drew attention to an unusual feature of the dance of the White-throated Manakin. At the point in the dance where other male manakins would point their tails in the air, a male of this species pointed its bill in the air. The replacement of tail-pointing with bill-pointing was unlikely to have happened as a better way of drawing attention to genetic quality of the male, he argued, because then it would have happened also in other manakin species. The scientists who reviewed his paper objected to this argument. He must prove, they said, that bill-pointing was not a better advertisement of quality in this one species. This was when the penny dropped for Prum that the good-genes argument had somehow become the null hypothesis and that Fisher's alternative of random, arbitrary elaboration had become the idea that needed to prove itself. The reviewers insisted that the burden of proof was on Prum to show that the switch had been arbitrary. But how can you show that something is without rhyme or reason? 'The prevailing standard of evidence meant it would be impossible for me to ever conclude that any trait had evolved to be arbitrarily beautiful. It had actually become impossible to be a contemporary Darwinian.' Prum went back to that forceful, if perhaps facetious, comment from Alan Grafen in 1990: 'To believe in the Fisher-Lande process as an explanation of sexual selection without abundant proof is methodologically wicked.' An irony is that the very concept of the null hypothesis in statistics was invented by none other than Fisher himself in a different context.

The reluctance of scientists to abandon an adaptive, rhyme-and-reason explanation for ornaments and dances in birds is perhaps part of the 'intentional stance' – the human tendency to believe that everything, from thunderstorms to bad luck to life itself, must have a specific, purposeful cause. We seem to have an aversion to causelessness: things happen for a reason. Prum went on to publish a paper in 2010 arguing that the Fisher-Lande-Kirkpatrick runaway hypothesis should be the null hypothesis. It 'predicts the evolution of arbitrary display traits that are neither honest nor dishonest, indicate nothing other than mating availability, and lack any meaning or design other than their potential to correspond to mating preferences'. To him, the very diversity and arbitrariness of sexual displays in nature demand a mechanism that is not bound by the need to advertise fitness. Sexual displays are seductive because they are seductive. Hence, as Prum was busy cataloguing in the forests of South America, the remarkable fact that there are fifty-four species of manakin, each with its own peculiar dance, and its own peculiar combination of bright male colours. The Red-capped Manakin, for instance, does a little backwards moonwalk

Club-winged Manakin displaying. Photo by Tim Laman.

along a branch; in the Lance-tailed Manakin pairs of males perform choreographed jumping displays from a branch.

Prum emphasises the astonishing lengths to which evolution seemed prepared to go in making males attractive to females. In 1985 he first came across a little-known bird called the Club-winged Manakin, hearing its bizarre 'wing-song' echo through an Ecuadorian cloud forest like 'feedback from an elfin electric guitar': 'Bip-Bip-WANNGG!' Three of the wing feathers of the male bird are contorted into strange shapes, with greatly thickened shafts and strange kinks and bumps, each feather different from the others. This, as Prum's student Kimberley Bostwick eventually showed with high-speed video, enabled the bird to make loud, tuneful 'songs' by rubbing the ridged feathers against each other at impossibly high speeds, in much the same way as a violin string vibrated when bowed. Here was a mechanical song device quite distinct from the true song the bird also made. Why bother? What possible rhyme or reason could lie behind a bird evolving a second, wing-singing mechanism to advertise the quality of a male? Bostwick went on to show that the needs of wing-song have led to the very bones of the bird evolving into radical new shapes. The ulna bone is roughly the same slender shape in all birds; in Club-winged Manakins it is thick, ridged and spoon-shaped – four times wider and three times the volume of other manakin ulnas – so that you would not recognise it as a wing bone at all in isolation. It is solid unlike every other bird ulna: even *Tyrannosaurus* ulnas (cousins of the earliest birds) are hollow. Little wonder then that the bird flies awkwardly. As Prum puts it: 'The Club-winged Manakin's wing song provides a likely stark example of evolutionary decadence – an evolved decrease in the overall survival capacity and fecundity of a population through mate choice.' And Prum finds that even the female of the species is lumbered with hefty ulnas poorly designed for efficient flight – which makes little sense from a good-genes version of sexual selection. A mating preference has driven the entire species down a hazardous evolutionary path. Prum thinks it is no accident that many of the 'most exquisitely beautiful and aesthetically extreme creatures are so rare', a point that has often occurred to me while watching the Black Grouse lek: is lekking a reason for the scarcity of the species?

'Beauty and desire are free to explore and to innovate'

In 2007 Prum broke new ground in the study of fossils, too, showing that you can tell the colours of the feathers on fossil dinosaurs from the shapes of their fossilised melanosomes, or pigment-carrying granules. Paleontologists had long assumed that the tiny sausage-shaped objects found among fossilised feathers were bacteria. Prum took one look and said they were more likely melanosomes, a conjecture confirmed when a Brazilian fossil of a black-striped feather from a hundred million years ago was put under the electron microscope and the sausages were found to be confined to the black stripes. The recent discovery from well-preserved Chinese fossils that many dinosaurs were feathered, like their bird cousins, then allowed Prum and his colleagues almost incredibly to reconstruct the colours of these creatures. Red, black, grey and white melanosomes are different shapes in modern birds. Prum mapped the shapes of melanosomes in the feathers of a small Jurassic theropod dinosaur called *Anchiornis huxleyi*. He unveiled the first coloured image of a feathered dinosaur. To my delight, it echoes a Black Grouse, having black and white striped wings and a bold red crest on the top of the head. It also has striped feathers sticking out to the rear of each leg, and a sprinkling of reddish feathers on the cheek.

It is highly unlikely that this pattern was camouflage: far too conspicuous. A bold, black and white body with a red head is – in grouse and other birds – an unmistakeable signature of sexual display. Because we now know that feathers adorned flightless dinosaurs for millions of years before any of them took to the air, Prum now thinks

that feathers were used for coloured display and subject to sexual selection long before they were used for flight. Female choice may therefore
have played a big role in the elaboration of life itself, because without
it feathers – which probably served as insulators to start with – may
never have grown long, diverse and elaborate enough to be used for
flight. In short, Prum's radical, ultra-Darwinian but anti-Wallacean
view goes well beyond the realm of traditional sexual selection theory
and puts female choice right at the heart of evolution itself. It is not a
side show. This notion is both bold and liberating: 'When mating preferences are unconstrained by the narrow task of providing adaptive
advantages, beauty and desire are free to explore and to innovate, and
thereby transform the natural world.'

A few weeks after my visit to Prum at Yale, while watching the Black
Grouse lek, it dawned on me that my species probably does not really
know the half of it about beauty. Not like the birds do and other dinosaurs did. They have been experimenting with bright colours for a
hundred million years. I'm a mammal and mammals don't do beauty
much. We mammals are almost all some shade of brown, and brown is
the default colour nature adopts when it is not trying to be 'colourful'.
We mammals almost never grow bright-coloured plumes or manes. We
grunt and shriek but we rarely sing, let alone as tunefully as birds.

Most mammals don't even do much colour vision and those of us
that do have a paltry three colour receptors, two of which (red and
green) are pretty close together in wavelength terms. Birds, insects, fish
– they all must think of us mammals as grim, dull, utilitarian, monochrome bores. Mantis shrimps have up to sixteen different colour
receptors in their eyes and a unique ability to see circularly polarised
light. Except in a few monkeys' faces, the smartest male mammals ever
get is to grow a pair of fancy horns or develop some bold stripes or
spots. Where is the bright red fur, the iridescent blue pelt, the golden
yellow skin, the green striped sides, even the ultra-black back? I suppose
it's because we – our ancestors – spent the Mesozoic era being mostly

'The male Great Snipe is perfectly camouflaged.'

'Darwin thought the bright colours and banded wings of the male Demoiselle are there to "charm" the female.'

'The Blue or Indian Peacock
is the paragon of sexual
selection theory.'

'Almost nothing is known about
the sex life of the Green Peafowl.'

(Photo by endri yana / Alamy Stock Photo)

'No two male Ruffs wear the same outfit.'

'There are fifty-four species of manakin,
each with its own peculiar dance.'

Wire-tailed Manakin.

(Photo by Matthew Dryden)

Red-capped Manakin.

(Photo by David Hill)

'The birds of paradise are sexual selection's masterpieces.'
– Raggiana (left) and Wilson's (right) Bird of Paradise. (Photo by Tim Laman)

'Birds are far more colourful than mammals – the Purple Grenadier Finch.'

'In the Golden Bowerbird's beak is a small, white flower that it places neatly.'

'Regent Bowerbirds construct flimsier bowers than Satin Bowerbirds.'

'He pops into the bower, does a little housework.'

'The male Great Bowerbird picks up a chilli pepper and proudly parades it while she stands in the bower.'

Darwin wrote that 'Birds have nearly the same taste for the beautiful as we have.'

'The effect is to show her every part of his plumage.'

'Mammals are prose; birds are poetry.'

nocturnal, while the dinosaur-birds ruled the day. Only after the aster-oid hit sixty-five million years ago, did we start slowly to become daylight lovers. So when we human beings wax lyrical about the beauty of a bird of paradise or a sunset, we are mere beginners, naive dullards glimpsing what the real gods of colour can do, and not appreciating it in its full glory. Indeed, the fact that birds can see ultraviolet colours is testimony to the fact that we mammals are truly ignorant of the glories of the world around us. Perhaps my obsession with birds is rooted in a form of envy. I wish I could see like a bird and grow feathers like a bird. And perhaps I need to slough off my mammalian obsession with rhyme and reason – rather than fun and show. Mammals are prose; birds are poetry.

9.

Paragon Peacock

Thus Argus lies in pieces, cold, and pale;
And all his hundred eyes, with all their light,
Are clos'd at once, in one perpetual night.
These Juno takes, that they no more may fail,
And spreads them in her Peacock's gaudy tail.

OVID, *Metamorphoses*

8.30 a.m., April, the farmhouse kitchen

Around 1980, when I was just starting out as a graduate student in evolutionary biology, it occurred to me that from Darwin onwards scientists had often cited the Blue or Indian Peacock as the paragon of sexual selection theory yet almost nothing was known about how Peacocks actually behaved. Nobody had done a proper study on how they used their massive trains or how Peahens responded to them, even in the semi-tame state in gardens in Europe, let alone in the wild in India where the birds lived naturally. This poster boy of sexual selection debates was a near-perfect blank when it came to actual evidence.

The bird books all said that a successful adult Peacock keeps a harem of females, implying that he follows a group of females persistently, jealously guarding his monopoly on mating with them. I knew about harem polygyny because it was the pattern used by another bird I was studying, the Common or Ring-necked Pheasant. I looked up various papers trying to find more details about Peafowl and drew a blank. It was as if the harem idea had been assumed rather than observed. Owning a harem is hard work, as Red Deer and Elephant Seals know: the male must exercise constant vigilance to protect the females from either murder by a predator or seduction by a rival male. Standing around looking fancy will not cut it, so why the big train of decorated feathers? Hence harem-defending males are usually large and armed with weapons, rather than colourful and decorated with flamboyant plumage. It just did not seem right that Peacocks were in this category.

Then I came across a brief remark in an article by the writer and artist Desmond Morris saying that far from chasing each other away, Peacocks seem to gather to display socially. I found another short note from the biologist Kumar Sharma, published in French in 1969, that said there was only a loose connection between the Peafowl sexes in the breeding season. This did not sound like harems. Perhaps conventional wisdom was wrong. I sat down and wrote an article for the *New Scientist* magazine entitled 'How the Peacock Got Its Tail', in which I emphasised the yawning gap between speculation and knowledge when it came to Peacocks. (Incidentally, the decorated feathers are not part of the tail, but so-called tail coverts on the lower back: hence I prefer the word 'train'.)

So it was that one fine, early spring day in 1982, as a break from analysing data, I went to an arboretum near Oxford where a small group of Peafowl lived freely among the rhododendron bushes and lawns. I sat down to eat my lunch and watched the birds. I returned the next day and it soon became a habit. There were four adult males with full trains, three young males with half a train or none, and seven females. It quickly became apparent that there was no sign of any mate guarding, let alone a harem. The males did no such work at all: they guarded, protected or monopolised the females not one bit. The females did their own thing all day, usually as a loose group, wandering freely and accompanied by the young males. In the afternoon the adult males sometimes joined them. But during the morning and evening, the males remained clumped loosely together, and each at his own separate spot, towards one end of the garden. Each male displayed frantically if a female came near but made no attempt to follow the female if she left, let alone corral or guard her. Towards each other, however, the males were aggressively territorial, each having his own patch within the clumped area where they spent their time. I suddenly realised I was looking at a lek: a concentration of fixed, small territories, used by males for display to free-roving females, the central, senior territories being relatively smaller than the peripheral, junior ones. I recruited two scientist colleagues, Mike Rands and Tony Lelliott, to take it in turns to help me record the exact positions of each male every day over the following weeks, and chronicle their behaviour: when they fought, when they displayed and when they mated.

Unsure whether our small group of birds was typical, I went to Whipsnade Zoo, where there was a free-living colony of Peafowl. Ignoring the lions and rhinos, I watched the Peafowl and sure enough I could see that the males aggregated in the morning in particular paddocks, each having his own small territory and display site. Later that year I found myself in western India during the monsoon, working on a conservation project on Lesser Floricans, and saw plentiful Peafowl around the grounds of the ancient palace near where I was staying. Being sacred in Hindu and Buddhist mythology, Peafowl are often very tame in India and live around villages, temples or palaces. Throughout much of the subcontinent, the species is almost semi-domesticated. Again, it looked like a lek to me. The males spent most of their time apart from the females, though clumped close to each other. The females visited their fixed territories, eliciting displays and sometimes mating. I encouraged a colleague, Nigella Hillgarth, a student of

Bill Hamilton's, to return to the same part of India the next year and do a short study of the Peafowl in the area. Again, she confirmed that Peacocks seemed to lek.

Ten years later, around Aligarh Fort in northern India, Shahla Yasmin and H.S.A. Yahya watched a population of wild (but people-friendly) Peafowl during their breeding season and recorded their mating success. Of the eleven males they watched, one achieved ten matings, two managed five, and one each had four, three and two matings; two had a single mating and another two had no matings. A typical lek pattern. The males that did best had the longest trains and the longest and most frequent calls. So there seems little doubt that

A Peacock lek at Whipsnade Zoo in the 1980s.

Peafowl do indeed vindicate Darwin: their astonishing plumage is the result of female selection at leks over many generations. But do they vindicate Fisher or Wallace as to why they choose thus?

Back in 1984 we wrote up our results and published a short paper for the journal *Animal Behaviour*, announcing the discovery that Peafowl are a lekking species, not a harem species. The world took little notice. While watching the birds, though, we noticed something obvious but that would have pleased Darwin: he was right about males displaying their ornaments in the presence of females. In the arboretum it was plain as a pikestaff that a Peacock spread his train into a fan at the approach of a Peahen and folded it away when she left. Indeed, I noticed that during the preferred display hours – early morning and late afternoon – when a male saw females approaching, he would make for one of his two or three display sites within his territory, often a sort of alcove among bushes or buildings, then spread his fan while turning away from the females. This has the effect of hiding the full glory of his display from an approaching female till the last minute. When she was close enough he would spin round to envelop her in the full technicolour radar-dish effect, shivering the entire screen of feathers as he did so. His behaviour plainly spoke of seduction: it had nothing to do with intimidating other males.

To turn observation into data, we recorded the birds' every move. It could not have been more obvious that the target of male display was the females. On nearly seven hundred occasions where hens were either approaching or walking away, a cock was twice as likely to spread his train into a fan as to fold it away when hens were approaching; and twice as likely to fold it away as to spread it when females were walking away. The train was never spread as a prelude to a face-off or fight with another male. When the male was on his display site with his fan already spread and a female approached, he ignored her just 24 times and turned suddenly towards her 651 times. On 573 of those occasions he then shivered the fan at her, making a strangely three-dimensional illusion where the eyes seemed to be suspended against a blurry, shimmering, iridescent background.

As for the typical female, early in the season her reaction to this flamboyant spectacle was usually to flap her wings once or twice. But when she was approaching readiness to mate, she reacted differently – running quickly round behind the male. And rather more forcefully than in Black Grouse, the males did eventually take the initiative and make a move. On seventy-eight occasions we watched as a male 'hootdashed' at a female (my name for the rather inelegant and noisy pounce the male would make). Mostly she then sprang away, seemingly unwilling, but on nine of these occasions we watched her squat and allow him to mate. The display was – unambiguously – a performance for an audience of one. None of this Peacock dance was in Darwin's book, more's the pity: why did he not bother to watch them when he was writing about them in the *Descent of Man*?

'Peahens prefer Peacocks with elaborate trains'

A few years later, when I was off doing other things, the biologist Marion Petrie – who had gained a fine reputation in evolutionary biology with an experimental study of Moorhens – was studying Chinese Water Deer at Whipsnade Zoo. She also spotted the possibility of observing the Peafowl there and told me of her plans to make some far more detailed observations. She teamed up with Tim Halliday and Carolyn Sanders of the Open University, with some help from Nigella Hillgarth again. Over the winter they caught 111 birds, about two-thirds of the population, and spent long days in an unheated hut with frozen fingers measuring every part of the birds' bodies: their weight and the length of their heads, beaks, legs, spurs, wings, tails and trains. They then continuously watched a single Peacock lek in a spot called Flint Pit paddock within the zoo. This lek had ten resident males. As we had observed, Petrie saw adult males use fixed display sites that were clumped closely together, while younger males floated more freely around the park – in typical lek fashion. In all she saw 291 hoot-dashes and 33 matings. Mating success was highly skewed: one male mated 12 times, two seven times each, two twice each, three once each, two not at all.

As with Black Grouse, females seemed to sample several males before deciding which one to mate with. Like Selous, Petrie found that – despite the hoot-dashing – it was clearly the female's decision as to who to mate with, and not a matter of males deciding for them: 'Our data do not suggest that competition between males is an important deter-minant of mating success amongst males established on leks. Whether or not a female chooses to mate with a particular male seems to be of overriding importance, regardless of the amount of sexual interference he is subject to.' As for what the females were choosing, it must be genes alone. Petrie wrote that 'Peacocks classically provide no resources other than seminal fluid to aid female reproduction. They do not fertil-ise more eggs per female, they do not courtship feed, they do not defend females from other males, they just stand displaying on the same small spot looking beautiful, and females are free to move between males.'

The most startling result in Petrie's study was announced in the title of her paper, published in 1991: 'Peahens Prefer Peacocks with Elaborate Trains'. By counting the eye spots on photographs of the trains of males they were able to tell which males had the most deco-rated trains. The successful cocks had more eye spots. Not only that, but when a female was observed to sample several males in succession she mated with the one with the most eye spots on ten of eleven occa-sions. Petrie was quick to point out that this did not prove the Peahens were counting eye spots as a cue to male quality: with around 170 spots, it's more likely that the females get a general impression. And in any case, the number of eye spots might simply correlate with something else about the bird. Yet, strikingly, they did find that none of the other measures of male quality – weight, wing length, spur length, etc. – mattered as much as eye-spot number: 'The only factor significantly correlated with mating success was the number of eye-spots in the male's train.'

Many years later, Jessica Yorzinski at the University of California Davis equipped some Peahens with a device called a miniaturised telemetric infrared gaze-tracker. This is a sort of plastic cap held in place by Velcro straps that includes a headset which measures precisely where the bird is looking. The bird wears a backpack containing batteries and a transmitter that projects what it is looking at to a nearby computer. Yorzinski found that the shaking of the train and the wings 'captured and maintained female attention' but that at close range the female looked intensely at the lower part of the fan-shaped train, not

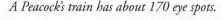

A Peacock's train has about 170 eye spots.

at the middle, and that she swung her gaze left and right along the bottom edge of the fan.

Back to the Whipsnade study. In the winter of 1989 Petrie and Halliday did the obvious experiment: they cut out some of the eye spots on some of the males' trains. They caught twenty-two males and on half of them removed twenty of the outermost eye spots with scissors. This did not alter the length of the train, because the longest feathers of all, the so-called fishtail feathers, have no eye spots. The result was that these males achieved on average 2.5 fewer matings than they had the year before, while a comparison group of males whose trains had been handled but not mutilated, saw no decline in mating success. So this seems to confirm that, as expected, females prefer ornate and intact males.

Petrie then did an ingenious and careful experiment. In February 1991 she captured eight of the Whipsnade Peacocks, including some that had been successful in mating the year before and some that had not, and took them into captivity. A month later she introduced into each of their eight cages four females bred in captivity, making sure that each male got – on average – females of the same size and weight. She then collected the eggs laid by these thirty-two females, 519 in total, hatched them under broody hens and reared them indoors under heat lamps. The 349 chicks that survived were repeatedly measured. A clear pattern emerged: the heaviest offspring at eighty-four days old were the ones with long-legged, long-trained or many-eye-spotted fathers. And especially larger eye spots: 'When all these variables were put into a multiple regression with other potentially confounding variables (hatch date and egg weight) only males with larger eye spots tended to have larger offspring.' This was true in both sexes of young and was independent of the weight of the eggs or the date of hatching.

She then released a matched sample of young Peafowl from each of the fathers into Whipsnade Zoo and observed which of them survived for a year while running the gauntlet of hungry foxes that patrolled the park at night. Again, it was the offspring of the males with the largest eye spots that were heavier and survived longer. Petrie concluded: 'The results show the offspring of highly ornamented males tend to grow better and that these advantages translate into differences in the chance of their subsequent survival under almost natural conditions.' To some extent at least, this supports the Wallacean, good-genes argument, that females are picking exaggerated ornaments because they are an indication of good genes: 'These data provide support for the idea that females may be gaining good "viability" genes for their offspring when mating with attractive males.'

But is it really because they are more ornamented that they do better? And whence comes their advantage – is it disease resistance or something else? Outstanding as Petrie's experiments and observations are, they do not rule out the possibility that it is the sexiness of the sons, rather than the health of the daughters (and sons) that motivates female choosiness at a lek. She did not record the eventual mating success of the offspring of preferred males but she did calculate how heritable train length was. By measuring various features of forty-two sires and eighty-six of their male offspring, she was able to detect that body weight, spur length and leg length showed little heritability, but train length was heritable to some extent. As Petrie pointed out, this was unexpected and seemed to challenge the lek paradox: 'This study adds to the building body of evidence that high levels of additive genetic variance can exist in secondary sexual traits under directional selection, but further emphasizes the main problem of what maintains this variation.'

If you were a Peacock in a zoological park in some parts of the world, the late twentieth century was a worrying time. After decades of being left in peace, you suddenly found yourself grabbed, probed, measured and bled, then eavesdropped upon. British and Indian Peafowl were now joined as scientific guinea pigs by French ones. In the early 2000s Adeline Loyau and her colleagues watched twenty-four Peacocks in the

Parc Zoologique de Clères in northern France and took blood samples from them. They found that the males with the smallest number of circulating antibodies in their blood displayed the most and got the most matings. There was only a weak association between eye-spot number and health, however. The twenty-four birds were then captured and injected with a solution of chemicals. Half the birds received a harmless saline injection. The other half received a solution of lipopoly-saccharides, which elicit a strong immune reaction. Over the following days, the ones with the immune challenge displayed less than the others, which was not surprising. More interestingly, the ones with the most eye spots on their tails were best able to cope with the challenge, reducing their display rate the least. Here at last was good evidence for the Grafen idea of a condition-dependent signal: that the cost of performing a sexual display should be higher in individuals in poor health. The display therefore may act as an honest indicator of the health status of the male bird.

The following year, Loyau performed another experiment, plucking some feathers from the trains of Peafowl and taking them to Bristol University for analysis of their colour and brightness. By laying the feathers on a black velvet background and measuring the hue, brightness and chroma from different angles, she and her collaborators were able to record just how 'iridescent' each was: how much its colour and brightness changed according to the angle from which it was seen. They concluded that the successful males had brighter eye spots whose colour contrast changed most when the angle changed. So maybe it was not the number and pattern of the eye spots but their individual iridescence that mattered most to the females.

'The Peacock's train is an obsolete signal'

Then, in 2008 a Japanese spanner was thrown into the works of the Peacock studies. This is the way science is supposed to work: rival teams challenge each other's results, with the conflict – however fractious – eventually helping to resolve any confirmation bias, motivated reasoning or plain chance that has infected the work and thus eventually settle on something approaching the truth. The spanner in this case belonged to Mariko Takahashi and her colleague Mariko Hiraiwa-Hasegawa, the latter a talented biologist who had begun her career studying Chimpanzees in Tanzania, followed by Fallow Deer (a lekking species) in Britain. In 1990 she began to watch free-ranging Peafowl in 1990 in Izu Shaboten (Cactus) Park near Ito, south of Tokyo.

Together with colleagues, Takahashi and Hasegawa closely observed fifty-two males and twenty-five females over seven years. They took particular notice of the 'run-around' response of a female to a male's display – which I had described and named a decade earlier. Run-arounds often preceded matings, they noticed, and so if a female was seen to run around a male three times they deemed that he was a 'preferred' male. With splendid precision they recorded that 'females on average met 11.2 territorial males 30.0 times (including 5.9 preferred visits), received 78.7 shivering bouts and accepted 0.9 copulations in response to 5.4 male-initiated copulatory attempts'. In all they witnessed 268 matings over seven years. Yet in their study neither the length of the train, nor the number of eye spots nor the symmetry of the train made much difference to a male's mating success. However, they did find that the successful males shivered their tails more often.

In an entertaining aside, Takahashi and Hasegawa spelled out frankly the contradictory results that had now been obtained by different studies of Peafowl: males that fell victim to predators had shorter trains than survivors – or longer trains; measures of the train were related to male ectoparasite loads – or were not; train length was related to male weight – or was not; the number of eye spots was related to male immunocompetence – or was not; and the most successful Peacocks were ones with longer trains – or without; and with more eye spots – or without. They concluded: 'Our findings indicate that the Peacock's train (1) is not currently the universal target of female choice, (2) shows small variance among males across populations, (3) does not appear to reliably reflect male condition and (4) is perhaps ancestral and static rather than recently derived.'

This last point was an intriguing and new one. What if female choice had indeed driven the enlargement of a Peacock's train in the past but had now stopped doing so? 'We propose that the Peacock's train is an obsolete signal for which female preference has already been lost or weakened, but which has nonetheless been maintained up to the present because it is required as a threshold cue to achieve stimulatory levels in females before mating.' Some time in the past Peacocks experienced a burst of evolution, driven by female choice, but it's over: females are still choosy but it makes little difference.

Loyau and Petrie soon responded, criticising Takahashi and Hasegawa for drawing such strong conclusions from a single negative study – after all, absence of evidence is not evidence of absence – and picking holes in some of their methods, as well as pointing out that the studies might still have missed whichever crucial feature of the plumage that mattered most. In particular, the Japanese Peafowl population was heavily male-biased, which would inevitably make it harder to pick up a pattern of female preference. In their response, Loyau and Petrie made an intriguing suggestion, one that offered a way to patch up the quarrel between Fisher and Grafen once and for all. The evidence, they conceded, 'provides strong support for the idea that the train evolved through Fisher's runaway process', but that was not the end of the story: 'At some point, as the train became an extreme trait, it presumably crossed a threshold and started to impose costs on the bearer, with only the better quality males being able to grow and display the more ornamented trains.' So the runaway stops working for lack of mutations and yet the species has stumbled on a good way to sort the feeble male chaff from the vigorous wheat. 'According to this scenario, the train has evolved as a Fisherian trait and is maintained as a good genes indicator, whatever the hormonal control.'

Thus they conceded Takahashi's and Hasegawa's intriguing suggestion that sexual selection might happen in bursts and then leave a species with an exaggerated ornament as a relic of the episode. They differed in whether the female's preference was now meaningful – as a way of maintaining a healthy lineage – or just a habit with no remaining purpose. Petrie and Loyau seemed to be agreeing with Takahashi and Hasegawa that the exaggeration of ornaments in the Peacock had happened in the past and had now reached some sort of impasse.

Next it was the turn of Canadian Peacocks to have their nuptial rites eavesdropped upon. Starting in 2007, Roslyn Dakin studied free-ranging Peafowl at Winnipeg Zoo, Toronto Zoo and the Los Angeles Arboretum. She counted the eye spots on each male's train and measured the length of the longest feathers in the train. Like Hasegawa, she found no relation between train length and mating success or between eye-spot number and mating success. Like Petrie, however, she did find that cutting twenty eye spots out of the train diminished a Peacock's average mating success. Dakin concluded that there is little natural variation in eye-spot number, at least in these inbred feral populations, for the female to discriminate among. Perhaps it is different in the wild in India. Every adult male grows about 165–170 eye-spot feathers but some are then lost in accidents. The lek paradox lives!

The Hamilton-Zuk theory that disease resistance might be what females are looking for when choosing a mate was eventually put to the test in Peafowl. In the late 1990s Marion Petrie teamed up with the Danish biologist Anders Møller to assess the immune systems of Peacocks using various procedures to test the immune reactions of those in captivity. They measured the concentration of white blood cells and immunoglobulins in different males and found that birds with larger trains had lower counts. But antibodies produced in response to injections of sheep blood cells and other measures of the reactivity of the immune system were positively correlated with train size. In other words, the males with the biggest trains were apparently more capable of fighting off parasites and less in need of doing so, just as Hamilton and Zuk would have predicted. However, neither immune measure correlated with the number and size of the eye spots, the feature Petrie had found Peahens paid attention to. Petrie thinks this may be because the males were two years old, and not yet equipped with a full quota of eye spots. So by choosing a long-trained male a Peahen might get a relatively good immune system for her offspring but choosing a many-eyed male – as she seemed inclined to do – may not help her achieve this.

Then, using a flock of Peacocks that had been breeding in captivity for many generations, Petrie and colleagues planned another, carefully controlled experiment. A total of forty-six males were penned separately with four randomly chosen females each for a whole season. Each male was genetically analysed to measure the diversity of its 'major histocompatibility complex' (MHC), a set of highly variable genes that effectively act as a sort of library for the immune system, giving it many different options for attacking germs. The males with more diverse MHC genes indeed proved to have stronger T-cell immune responses to challenge with antigens. They also had longer trains. So choosing a longer-trained male would appear to be a way for a female to get good genes for her offspring. Petrie and her colleagues then found that females with higher MHC diversity tended to lay more and larger eggs. Also, females with such high genetic diversity laid larger eggs for the males that had a similarly higher diversity, but females with low diversity did not lay significantly larger eggs for males with high diversity. 'Our results suggest that MHC diversity is a quality of potential mates that females assess and respond to via increased reproductive investment,' they concluded. And since males with higher MHC diversity had longer trains, and previous studies had shown a female preference for longer trains, it looked like a slam dunk that female choice for long trains was the means by which females were picking good genes. Frustratingly, however, in this study no such effect emerged: 'In the current study, train length was not a significant predictor of either egg weight or the number of eggs laid.'

'The male that most strongly excites her sexual instinct'

The Peacock studies have generally seemed to support the idea that disease resistance is what female birds are after when they choose mates based on display, but not conclusively. Like Boyce's Sage Grouse studies, they cannot rule out the alternative, Fisher-based hypothesis, namely that the main cause of the exaggeration of the Peacock's train was females following a fashion driven by the need to have sexy sons, rather than because they got better genes by being choosy.

Forty years of studies on Peafowl mating habits, since I sat down with my sandwich at an arboretum in Oxfordshire in 1982, have not yet delivered a final verdict. They have confirmed that Peacocks lek, that females choose which males to mate with, the males having little ability to force the matter, but they have not clarified why the females are so choosy. Is the Peacock's train just an obsolete relic of a past burst of sexual selection? Or is it a vivid and revealing indicator of the male's current disease resistance? Was it first one and then the other? We don't know. I love scientific mysteries.

In 2006, Marion Petrie raised another possibility altogether. She went back to the lek paradox, the notion that if a few males get most of the matings in generation after generation, there will soon be little to choose between them and therefore no point in being choosy. It sure does not look like that is what has happened, she argued, because there is still a big difference between a strong healthy male who displays a lot and a weaker one with a shorter train who dances less. Where is the variation coming from? Perhaps the answer lies in the genes. Might there be a genetic mutation in Peafowl, and other lekking species, that accelerates the rate of natural mutation elsewhere in the genome of the bird, generating fresh variability for female choosiness to feast upon? Teaming up with the theoretician Gilbert Roberts, Petrie set up a model inside a computer, simulating a population of animals in which some individuals had a 'mutator' gene that randomly altered other genes. Over twenty thousand generations, when the animals mated randomly, the mutator gene was disadvantageous because it more often caused a debility than an advantage, so the mutator itself eventually went extinct. In a population of animals practising female choice, however, where only a few males were chosen in each generation, the mutator gene thrived and spread. Its capacity for generating random variation was turned to advantage by the female's discrimination in favour of fitter males. This was true even when the females had a choice of just two males. A gene for increased mutation and a gene for female choice reinforced each other. 'Our results suggest that all that is necessary for the maintenance of an increased mutation rate (which is under genetic control) is a genetic benefit that outweighs the genetic costs of an increase in mutation rate. Female choice for good genes in our model provides that genetic benefit.' In summary, there needs to be variation among males for female choice to serve a purpose, but there needs to be female choice for variation among males to evolve and persist. It's a highly intriguing idea and a positive feedback loop of the very kind that could run away with itself.

In these open-minded days when even *Science* magazine 'highlights the value of teaching Indigenous knowledge alongside science in the classroom', I feel obliged to point out that there is a rival explanation for the Peacock's train. In Greek mythology, the Peacock was said to have been spawned from the corpse of Argos Panoptes, a giant with one hundred eyes. Argos, the all-seeing hulk, was employed by the goddess Hera for the usual sorts of chores you need help with on Olympus: killing a serpent-legged monster and guarding a white heifer in case her (Hera's) boyfriend Zeus tried to rape it – because it's not really a cow, you see, but the nymph Io in disguise. Argos was good at this security-guard job because not all of his eyes needed to sleep at once. But not good enough because Zeus simply got Hermes to kill Argos, a feat he achieved by either hypnotising the eyes one by one with spells, or by throwing a rock at him: take your pick from different sources. Either way, Hermes cut the awful giant's eyeful head off. Hera, mourning Argos (while watching the heifer Io cavorting with her husband before swimming to Egypt, as you do), decided to turn the body of Argos into a Peacock. Or perhaps (recollections vary) she just plucked out the eyes and plonked them on a Peacock's train. Christian mythology has just as bonkers an explanation of the Peacock's tail: the seven deadly sins complained to God that they were colourless so God put the colours on the tail of the Peacock, which now trails the sins behind it.

The smidgen of truth in this crazy tale – and no, there is never anything remotely 'valid' about such mythology – is that the decorations on a Peacock's train are not mere random blobs. They really do look like eyes, with their concentric colours and iridescent shine. Is this an accident or is there a reason that eye-like ocelli crop up also in the Argus Pheasant and seven out of the eight species of Peacock-pheasant? Eye-like spots are common motifs in many animals, including caterpillars, butterflies, moths and fish. In those cases they act as frighteners, by giving the impression that there's an owl or a snake staring at whichever predator is about to eat the prey in question. The Peacock butterfly is an example. Eyes catch the eye. As the art historian Ernst Gombrich stressed, artists have known for centuries that the viewer's gaze is drawn to the eyes of a subject, and the eyes seem to follow the viewer. Little wonder that Argus Pheasants place their real eyes at the centre of their displays. Likewise, in two species of pheasants with gaudily painted plumage, the Golden Pheasant and Lady Amherst's Pheasant, the male opens a ruff of concentric patterns around his real eye when he displays, shrinking the pupil of the eye as he does so.

In having a display that features hundreds of exaggerated eye-like patterns, the Peacock is in effect exploiting a pre-existing attention bias in the Peahen. Maybe, I once argued, the effect on the Peahen is to temporarily hypnotise her with staring eyes, and in doing so cause her

A Peacock putting on a show in a theatre garden.

to stand still long enough to allow mating. The better the display, the stronger the hypnosis. This revives Karl Groos's idea that 'a kind of unconscious choosing does take place which is in a peculiar sense sexual selection, for the female is undoubtedly more easily won by the male that most strongly excites her sexual instinct'. Briefly in the 1990s, this led to a new explanation for sexual selection, advanced by a Texas-based scientist named Michael Ryan. He noted that many female animals are attracted to exaggerated versions of displays that are beyond the range available in the local male population. Some female frogs were more stimulated by and attracted to the calls of males from a foreign population than any of the calls of the local population, for instance. He went one further and studied four species of frog, two of which have added a 'chuck' call to their whine and showed that the chuck call is especially resonant in the female's ear and important in stimulating her. But the two species without the chuck call have similar sensitivity to the wavelengths of sound in the call. Their males would do well to consider adding a 'chuck' call. The preference was there before the stimulus. Likewise, the attention-grabbing quality of eye-like spots probably existed in ancestral Peahens and female Argus Pheasants before their males ever evolved eye spots. In my mind, however, this sensory-bias theory explains what features will be exaggerated by female choice rather than how the exaggeration itself evolves. That remains a battle between Fisher and Wallace.

'The superior adaptability of the blue peafowl over green peafowl'

So far I have written only about one species of Peafowl, the one that lives all around the world in parks and gardens. This is the Indian, or Blue Peafowl, whose native range is throughout the Indian subcontinent. But there is another Peafowl, the Green Peafowl, which lives further east in south-east Asia. This is a long-legged, long-necked, slimmer bird, similar in weight and adorned with an almost identical train, which it spreads into a fan in the same way when displaying. But although the eye-bedecked train is much the same, the rest of the body is markedly different. The neck is green, rather than blue, and the feathers have a scaley appearance thanks to fine dark margins. The back is bright blue rather than patterned grey. The primary wing feathers are cream-coloured rather than reddish brown. The face is pale blue rather than grey, the chin is yellow rather than blue and the crest is a very different shape. In short, Mother Nature has decided to make every part of the bird different apart from the feature it displays during mating – which is the very opposite of what you might expect, given how the ornaments are the first bits to diverge in closely related lekking species such as grouse or birds of paradise. Yet here is a species in which males look different in every respect except for the uniquely extravagant ornament. Most peculiar of all, female Green Peafowl look exactly like males except that they lack the train. Female Blue Peafowl are dully grey birds, well camouflaged on the nest.

Bizarrely, almost nothing is known about the sex life of the Green Peafowl. The bird is rare, wary and unable to cope with cold weather so cannot survive in northern countries except in indoor conditions in zoos. I once spent a week in a remote, dry forest in Thailand hoping to watch Green Peafowl but was almost entirely defeated. We heard them calling, but they were far too wary to allow us to catch more than occasional glimpses. We found their tracks in the sand along the banks of a small river that ran through the reserve and hit upon the idea of wiping the sand clear of footprints so we could return the next day and see which places they had visited to drink from the river. This worked but caused us some alarm when one morning there were fresh tiger tracks on the sand. One day I tried sitting up an isolated tree in a forest clearing in the hope that the birds would walk past but all that came along – to my consternation given that I was fairly conspicuous – was a group of camouflaged soldiers carrying automatic rifles. It was a Thai army border patrol and they were much amused, and a little bit suspicious, to find a lanky Brit up a tree.

Judging from more recent accounts from around south-east Asia, including one from the same nature reserve where I saw them, Green Peafowl seem to travel in small flocks, often near rivers, and it appears that in the breeding season males display singly at particular spots. Nobody seems to have come across anything resembling a lek. But a piece of intriguing genetic evidence suggests that there might in fact be less polygyny in Green Peafowl than there is in Blue Peafowl. It relates to the diversity of gene sequences on the sex chromosomes. In mammals, females have two similar sex chromosomes called XX, whereas males have two different sex chromosomes called XY. In birds it is the other way round: males have two similar chromosomes, called ZZ, and females have two different ones, called ZW. This means that Z chromosomes spend twice as much time in males as in females. If males have highly skewed mating success, with some individuals mating many times and others not at all, then Z chromosomes should come to have less genetic diversity. They always show less genetic diversity than other chromosomes since they spend half their time as single copies (that is, in females). But sexual selection should lower this 'Z/A' ratio of genetic diversity even further. A study of shorebirds found general support for this hypothesis: the more polygynous the species, the lower the ratio. Intriguingly, a genetic analysis done at the Indian Institute of Science Education and Research in Bhopal and published in 2022 found a much lower Z/A ratio in Blue Peafowl than in Green Peafowl. This, the authors suggested, might indicate that the Green Peafowl is actually monogamous, or certainly less polygynous than the Blue.

Another team at the same Bhopal institute found something even more striking about the genomes of the two species. Blue Peafowl have more genes, and by a huge margin. Their analysis found nearly 26,000 coding genes in Blue Peafowl, just under 15,000 in Green Peafowl. The difference, they think, is due to more duplicated genes in the Blue Peafowl in which different segments of each gene have diverged, as well as more gene clusters. They put this down to the fact that Green Peafowl seem to have gone through a genetic bottleneck around twenty thousand years ago when the population was very small. At the peak of the last ice age, when the world was much drier than today, south-east Asian forests would have shrunk, isolating Green Peafowl in small refuges of habitat. Blue Peafowl, adapted already to the drier wood-lands of India and then adopted as a friend by human beings, maintained a larger population, which allowed for more genetic diver-sity, especially in genes relating to neural development and other systems necessary for adaptability. The diverse genes include ones for opsins, used in vision, perhaps hinting at greater sensitivity to sexual displays and visual discrimination in the Blue Peafowl. The Bhopal discovery has implications for conservation, with the endangered Green Peafowl struggling to adapt fast enough to changes in its habitat caused by human activity: 'These genomic insights obtained from the high-quality genome assembly of *P. cristatus* constructed in this study provide new clues on the superior adaptability of the blue peafowl over green peafowl despite having a recent species divergence time.' Increasingly, it seems that to understand the behaviour of a bird, you need to peek inside its genes. This will prove even more true of the next flamboyant bird I watch.

10.

The Riddle of the Ruff

Did not the heavenly rhetoric of thine eye,
'Gainst whom the world cannot hold argument,
Persuade my heart to this false perjury?
Vows for thee broke deserve not punishment.
A woman I forswore, but I will prove,
Thou being a goddess, I forswore not thee.

WILLIAM SHAKESPEARE, *Love's Labour's Lost*

Midnight, midsummer, Arctic Norway

It is early June and I am sitting on a crowberry bank above a sedge marsh next to the Barents Sea in the far north of Arctic Norway, looking east towards Russia. Some domesticated reindeer are chewing their cud nearby and a massive Sea Eagle just flew low along the sandy shore. A couple of Common Snipe are drumming above the marsh, just as they do above the Black Grouse lek at home. It is midnight but I am five hundred miles further north than the Great Snipe lek so the sun is shining, its rays beaming in low from above the Pacific Ocean, passing over the North Pole itself. A keen breeze off the snowy plateau to the west keeps the temperature chilly. Feeding in the sedge marsh in front of me is a small flock of about fifteen multicoloured birds, each one different. They are clearly friends, the flock moving about together despite them being all males and this being the middle of the breeding season. That's because, once again, this is a lekking species and as I keep being reminded, lekking is a male team sport in which fierce rivals are also close comrades, dear enemies.

The birds are Ruffs and I am waiting for them to fly on to a low ridge of crewcut crowberry tundra to begin lekking. I have come this far north not just to see this most complicated of all lekking birds (it was once common in Britain but is now extremely rare). I am here also because this species breaks a basic rule of biology. Unlike every other species of bird, indeed of animal, there is an astonishing individuality about the Ruff: no two males wear the same outfit. Biology, somebody once said, is the science of exceptions rather than rules, and here is a very puzzling one. I need to tangle with this enigma.

A 'ruff' of long, full feathers around his neck and a pair of 'head tufts'.

Of course, if you look closely you can find small differences between individuals in all species. But these are blemishes or differences of degree not kind: bigger black bibs on some house sparrows than others, for example, not huge, obvious, opposite and colourful differences in every individual. Each male Ruff (the bird) has a 'ruff' (the ornament) of long, full feathers around his neck and a pair of 'head tufts' of feathers that can be erected behind the head. The ruff might be brown or orange-red, or tawny, or black, or dark blue, or buff, or bright white, or cream, or grey. It might be plain, or multicoloured, or flecked, or striped, or patterned. The head tufts can be any of these colours and patterns too, giving an infinite variety to the birds. They are as variable as a clothing catalogue or a room full of (female) wedding guests keen not to be seen in the same outfits.

Then there's the facial wattles: Ruffs lose the feathers on their face in the spring, growing small pimply warts or wattles around their eyes and beaks instead. These wattles can be yellow, red or orange. The feathers on the birds' backs are also coloured variably: sometimes reddish brown, sometimes grey, sometimes partly white. In the flock I am watching, there are birds with orange ruffs, blue-black ruffs, pale ruffs, white ruffs, striped ruffs, patterned ruffs. Almost any combination can occur, although apparently some combinations such as rusty ruffs with white head tufts are generally not seen. But the key point is that every bird is different. According to Georges-Louis Leclerc, the Comte de Buffon, writing in the 1770s, a naturalist named Klien 'compared above a hundred Ruffs together and found only two that were similar'.

This is the fourth flock of lekking Ruffs I have watched today and this variability is the same in each case: every flock wears a selection of different-coloured ruffs, almost as if each has carefully chosen to include one of each colour and kind. I cannot think of another bird – another animal – in which this is the case. In feral pigeons in cities, there is great variety of plumage, but that is largely human made:

Males grow pimply warts or wattles on the face.

people, including Darwin, have been deliberately breeding plumage variety into tame rock doves for a couple of centuries and the birds have spread their varied genes to their feral cousins. Similar human-made individuality is seen in cats.

In some wild species of birds such as Paradise Flycatchers and Arctic Skuas there are two or more different colour 'morphs'. But to find that every male is different, every male unique, in bold, colourful ways, in a wild species – there is no parallel for this. Only a strange, hyper-social bird in Africa called the Red-billed Quelea comes even close to achieving this individuality but it's a lot less varied than the Ruff. There are more than eleven thousand other recognised species of bird; none of them is like this.

During the winter, when the Ruffs are in Africa, this variety largely vanishes as the birds shed their ruffs and look much like slightly larger versions of their females, known as Reeves. But come the spring, as the birds migrate north from Africa to northern Europe and Russia, their ruffs and head tufts grow thick and long and different in every case. It's a puzzle – for if it is of use in the Ruff species to be individually different, why not in other species? And what is it about the ecology of the Ruff that explains this peculiarity?

The short answer: nothing. The Ruff is a sandpiper, one of hundreds of mud-loving species of wader or shorebird. It's big for a sandpiper, but medium-sized for a shorebird generally: larger than a snipe, smaller than a Curlew. It has a medium-sized beak: longer than a plover's, shorter than a godwit's; more curved than a Redshank's, less curved than a Curlew's. Its legs are longer than a Sanderling's but shorter than an Avocet's. It has a long migration but not the longest. Like many other sandpipers it frequents muddy places in the winter, Arctic or coastal marshes in the summer. It's not far off the archetypal shorebird. Apart from the fact that it leks, grows fancy plumage and every male looks unique, it is as average and boring a bird as you can imagine. So why the heck has it been subject to an extreme and aberrant burst of sexual selection, producing infinite individual differentiation?

I'm here because the Ruff seems to be a living rebuke to all the theories I have been discussing so far. Remember, in lekking species, I concur with Darwin that female choice results in extravagant plumage because females all agree on who the sexiest male is in each generation, so female selectivity drives progressive exaggeration in plumage and display – but also progressive homogenisation. Whether the females agree mainly because they want sexy sons or healthy daughters is unclear still but that they agree is pretty well certain. And the consequence is males that grow exactly the same plumage in every generation throughout the range of the species and differ only in fine distinctions. A Black Grouse in eastern Siberia has the same plumage, the same rituals and the same calls as a Black Grouse in Britain. I have watched videos of Black Grouse leks from northern China to compare and found the calls, the moves and the plumage indistinguishable from those I know. Yet here is a species in which females are just as selective – one male gets most of the matings on a Ruff lek – but the males do not look alike, even on the same lek, let alone in different regions. How can this be? The Ruff paradox is that the lek paradox apparently does not apply to Ruffs.

'Her caprices cannot be overruled'

Edmund Selous watched Ruff leks in April and May 1906, before he watched Black Grouse. He spent almost every dawn for several weeks crouching in a turf shelter in a marsh in Holland, there to test Darwin's theory that female choice happens. Being idiosyncratic he had not fallen for the fashionable Wallacean dogma that Darwin was wrong. After weeks of cold early mornings observing a Ruff lek, Selous had seen dozens of matings and his conclusion could not have been more clear. The lek was a 'courting place not the [jousting] lists'. Love not war was the business being transacted. Again and again he watched Reeves wander through the lek unmolested by males, pick one male and in unmistakeable fashion solicit what he euphemistically called the 'nuptial rite'. Most of the females chose the same male.

What Selous pointed out for the first time was that the females were very much in charge of what happened. A Reeve, he wrote, is 'not to be compelled, she is not to be intimidated, her caprices cannot be over-ruled'. The male appears to know this, which is presumably why through evolution by natural or sexual selection his forebears have come to place 'more and more reliance [on] pleasing the female, and less and less upon the likelihood of winning her by force'. Nor was the winning of jousts by the strongest male what seemed to matter to the

Reeves wander through the lek unmolested by males.

females. Selous saw no evidence of Reeves being impressed by the fighting qualities of males, but he did see 'strong and sustained evidence' that the Reeve 'has the full power of choice, and that she exercises it'. He concluded that his observations strongly supported Darwin's 'second great hypothesis of sexual selection'.

At first glance a Ruff lek is just like a Black Grouse lek. Just as in Black Grouse (and Great Snipe) there are about ten to fifteen birds on a typical lek, each with his own small court or territory. Just as in Black Grouse, young male Ruffs (known as marginals) hang around several leks trying to join in and get regularly beaten up for their pains. Just as in Black Grouse, fighting breaks out frequently among territorial birds and can sometimes get vicious. Just as in Black Grouse, the approach of a female causes all the males to fluff out their ruffs, erect their head tufts and go through a series of ritualised postures, bowing and standing tall, fluttering their wings and generally showing off. Just as in Black Grouse, Reeves visit the lek on several occasions over the course of a week or so before mostly selecting the same central male as each other and soliciting mating from him.

However, there are a couple of strikingly different features. First, the lek is silent. There is no singing, sneezing, crooning, clicking or calling of any kind. Just a gentle fluttering of wings in display and frantic thumping of wings in a fight. (The Ruff biologist Dr David Lank thinks this may explain the variability of colours: most birds recognise individuals by their calls; Ruffs for some reason have switched to visual individuality.) Second, the females do not confine their interest to a single male. Unlike most lekking species they make a point of mating several times with at least two males. They might choose to do this on two different leks, or to come back and choose a different male on the same lek. So females are choosy but they are also apparently careful to spread their bets.

But there's another surprise. Back in the 1960s close observers of Ruff leks in the Netherlands began to detect a strange pattern. The males with very pale or bright white ruffs and head tufts behaved differently from those with reddish, grey or dark plumage. These birds, which for some reason they unhelpfully called 'satellite' males, it gradually became clear, do not hold territories but move freely about the arena, largely unchallenged by the darker-ruffed territorial males (known as 'independents') and they display wherever they feel like. The independents often tolerate the satellites on their territories, and perhaps even encourage their presence in order to help draw in females. You sometimes see a dark-ruffed independent male and a white-ruffed satellite male bowing together with their ruffs spread and holding this frozen position facing each other inches apart in a coordinated display when a female is nearby. The independent will not tolerate another male in his court unless it has a white ruff. The paired display seems to be especially attractive to females. And it seems to be the middle-ranking independents, rather than the dominants, which co-display with satellites in this way, so the effect of the satellites is to even the odds of mating among the independents.

Exactly who gets to do the mating when a duo attracts an interested female is sometimes then a matter of fierce dispute between the two birds, but things can happen fast, distractions are common and the territory owner does not always prevent the satellite from mating. In general, very dominant independents do not let satellites mate, so satellites tend to co-display in the territories of second-rank independents, boosting the latter's attractiveness to females. These second-rank independents are more likely to welcome the satellites into their territories, accepting the risk of the satellite mating, in order to increase their chances of seducing a female at all. It's a mighty complicated game.

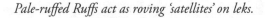

Pale-ruffed Ruffs act as roving 'satellites' on leks.

'It took a Friesian potato farmer to realise what's actually going on'

In the summer of 1983, a young, independent-minded bird researcher with a 'ruff' of curly hair turned up in Europe with a plan to study Ruffs in a new way. David 'Dov' Lank had experience capturing Arctic sandpipers in Canada and keeping them in aviaries to study their migration instincts. He wanted to keep Ruffs in captivity and study their sex lives up close. At Oulu in Finland, he found what he wanted in a dense population of breeding Ruffs and some willing local farmers prepared to help him catch and keep the birds. By the summer of 1985 he had persuaded Finnair to let him put egg incubators in passenger seats and fly 40 eggs to Helsinki and on to Montreal. There 32 of the eggs hatched, 16 of each sex. He raised them on a diet of poultry crumbs and established the only colony of Ruffs in North America, in large outdoor aviaries on a farm outside Kingston, Ontario.

Lank lights up with enthusiasm as he recounts to me how he became an accomplished 'Ruff farmer' over the next few years. He maintained his Ruff zoo for thirty-five years, latterly at Simon Fraser University in British Columbia, before finally dispersing the birds to a lab in Germany during the pandemic in 2021. At one point he had four hundred birds. This enabled him to control the paternity of each chick that was born in his little zoo (by offering each female only one male) and so find out what determines whether a young Ruff becomes a satellite or an independent. It was nature not nurture, he quickly found, and the cause proved to be surprisingly simple and implacably deterministic. Independents nearly always fathered independents but satellites fathered both satellites and independents in roughly equal numbers. This, together with back-crossing, soon strongly implied a simple Mendelian genetic pattern of exactly the kind first discerned by that inquisitive abbot with pea plants in a Czech monastery garden in the 1860s. Which meant there must be a single genetic factor involved.

Surprisingly, Lank then found that the gene was apparently not on one of the birds' sex chromosomes, as a sex-linked feature might often be, but on one of the ordinary chromosomes. It seemed that the Ruffs have a gene responsible for their ruff colour and choice of strategy, the 'independent' version of which is recessive and the 'satellite' version is dominant. Thus, to clarify: each bird would inherit two copies of the gene, one from each parent. Obviously, if they are both the 'independent' version of the gene, the bird becomes an independent male; but if a bird inherits one of each, then it will be a satellite – that's what dominant means in the Mendelian system. But if an embryo inherits two copies of the 'satellite' mutation it never develops so there are no birds with two copies of that mutation. The 15 per cent of males that behave like satellites are actually 'heterozygotes', looking and behaving as white-ruffed satellites but silently carrying the dark-ruffed independent gene.

To make matters even more complicated, in 2004 an observant Dutch farmer spotted something that led to another intriguing genetic discovery. Joop Jukema farms potatoes in Friesland in the northern Netherlands, in an area where netting shorebirds for sale as food was an ancient tradition. He is skilled in operating a 'wilsternet', a large screen net which, if sprung at exactly the right moment, can trap a flock of shorebirds lured by decoys and whistled bird calls. The practice of netting such birds for food was outlawed in the 1970s but at some point Dutch scientists realised they could draw upon the skills of these expert 'wilsterflappers' to catch birds for study instead.

The background here is that Ruffs were 'highly esteemed as a most delicious dish' (wrote Thomas Bewick in 1800) and were caught in huge numbers 'for the markets of the metropolis'. In 1466 some two hundred dozen of the birds were supplied for a single banquet at Cawood in Yorkshire. Before being sold the birds were fattened on bread and milk for two weeks, the 'feeder' making a handsome profit by charging three times as much as he paid the 'fowler' for catching them. In the 1760s one Ruff catcher in Lincolnshire boasted of having set off on a single journey to Ireland to supply his wealthy customers with twenty-seven dozen of the birds. By the late 1800s this trade in Ruffs in Britain and the Netherlands had driven the species to the brink of extinction in both countries. But Scandinavian birds still stopped off in good numbers in the Low Countries on their way north from Africa and it was here that the wilsterflappers intercepted them.

One day in 2004 Jukema was helping to net migrating Ruffs for Dr Theunis Piersma of the University of Groningen when he noticed something odd about one of the birds he had caught. It looked like a female, lacking a ruff, but was bigger, almost as large as a typical male. Piersma was dismissive at first of Jukema's suggestion that this was a male Ruff in female plumage, but from the 1,134 Ruffs caught that year, there were nine such birds and DNA tests soon proved that Jukema was right. These birds are male; indeed, they have even bigger testes than normal Ruffs, despite being slightly smaller in the body. 'All these scientists had been staring at Ruffs for years and it took a Friesian potato farmer to realise what's actually going on,' says Lank. In fairness there had been occasional reports of 'naked naped males' since the 1930s but they had been ignored.

Jukema coined the name 'faeder' for these ruffless males. Faeders are rare – only about 1 per cent of males – and we now know they have a cunning sexual strategy. They can move freely through the lek, experiencing an occasional homosexual copulation from independents, but then, presumably to the general surprise of both sexes, quickly pouncing when a female squats to solicit mating from a displaying independent. The independent male seemingly fails to realise that this apparently lesbian event has just pre-empted him. Shortly after Jukema's discovery, one of the faeders he caught was photographed on a lek very close to where I am now sitting, a thousand miles to the north of the Netherlands, and was seen to mate with both males and females. In winter, faeders flock with other males, not females.

Such 'sneaky' techniques are not unknown in other species, being adopted often by young males. Tiny salmon parr, for instance, sneak in and fertilise some of the eggs of giant females, undeterred by the presence of giant males. But the case of the Ruff is different. Faeders are not young male birds, but genetically distinct male types, fixed that way from birth till death.

The search was now on to find the gene involved in making both faeders and satellites. Next-generation DNA sequencing machines had collapsed the cost and exploded the productivity of genomic sequencing so Lank teamed up with Terry Burke of Sheffield University who recruited Clemens Küpper of the Max Planck Institute in Seewiesen in Germany to help him do the gene-finding. They then found themselves in a scientific race with another team based mainly at Uppsala University in Sweden on the same trail. Both teams succeeded in simultaneously mapping the genome of the Ruff and quickly located the gene responsible for deciding whether a male was a satellite, independent or faeder. Both agreed it was on a particular section of chromosome 11, instantly confirming each other's results. They published simultaneously in November 2015.

But there was a surprise in store. The relevant segment of genetic code proved not to be a single gene but a rearrangement of a lot of genes. A 'supergene' section of DNA, about 4.5 million base-pairs in length, and containing more than 120 genes, has been inverted in the faeder, so it reads backwards compared with the independent version. This inversion first appeared around 3.9 million years ago, according to molecular-clock calculations that count the random point mutations inside the supergene.

Much more recently, around seventy thousand years ago, there was another rearrangement of the same supergene on chromosome 11. This time what happened was a rare double-crossover event, in which parts

of the supergene recombined between a faeder genome and an independent genome in a bird that had inherited both versions of the
chromosome, so that some birds ended up with parts of the supergene
reading forwards and parts reading backwards. The birds that carry one
copy of this new mutation, which are true hybrids between faeders and
independents, are the satellite males with their white ruffs and roving
tactics. It is also now clear why satellite-satellite, faeder-faeder and
satellite-faeder genotypes cannot exist. Both mutations disrupt a crucial
gene called CENPEN, without which cell division is impossible.

This means that the faeder and satellite inversions only ever exist in
single copies. This in turn means they cannot recombine with their
partner chromosome, which in turn means they cannot be checked for
mutations and corrected. They are thus prone to accumulating mistakes
and sure enough Küpper finds that both inversions show signs of
advancing degeneration: 'A large number of deletions, insertions and
duplications of segments and missense mutations within the inversion

Most Ruffs have dark ruffs.

region on both the Satellite and Faeder haplotypes point towards gradual erosion of the Faeder and Satellite variants.' But Küpper argues that intriguingly this degeneration, while increasingly harmful to female survival, may be helpful to males: it has lowered the tendency to aggression and courtship, the better to enable them to do their sneaky matings on the lek. This is a phenomenon known to geneticists as sexual antagonism, where a gene benefits one sex and hurts the other.

Küpper now thinks he knows exactly how the rearrangement of the genes results in differences in concentrations of key hormones associated with different behaviours. One of the genes in the inversion region, called HSD17B2, plays a key role in converting testosterone to androstenedione. The inversion has altered the 'expression' of this gene, such that it is more active, which results in faeders and satellites turning much of their testosterone into androstenedione, a milder hormone that makes them less aggressive. He and his colleagues have now shown that the more expressive gene does indeed result in lower testosterone levels in faeders and satellites.

'The phenotypic gambit fails'

Given that there are three different morphs on Ruff leks, why don't two of them go extinct? Surely one tactic works better than the others? Not necessarily. To understand why, consider the children's game of rock, paper, scissors. Paper beats rock by wrapping it up, scissors beat paper by cutting it up and rock beats scissors by breaking them. So long as the players play the strategies equally often and randomly, there is no winning strategy. But if you play the game against somebody who tends to play rock most often, it makes sense to play paper; if he plays paper most often then it pays to play scissors; and if he plays scissors most often then it pays to play rock. The commonest strategy therefore has a disadvantage, so no strategy will go extinct.

The rock, paper, scissors game was adopted as an analogy by Barry Sinervo and Curt Lively of Indiana University to describe their study of side-blotched lizards in California in the 1990s. The throats of male lizards come in three genetically determined colours: orange, blue and speckled-yellow. Orange lizards have the biggest territories and the biggest harems; they can dominate blue lizards, invading their territories and stealing matings with their mates. Blue lizards can dominate yellow lizards, who look like females and do not even try to hold territories. But – and here's the twist – yellow lizards can sneak in and steal matings with the mates of orange lizards while the latter are off stealing blue lizards' mates. So yellows undermine the success of orange which undermine the success of blue which undermine the success of yellow. Rock-paper-scissors.

Perhaps this is also the explanation in Ruffs: if you are the only satellite male or the only faeder on a lek full of independents you will do well. But once there are lots of your kind about, your competitive advantage will fade. Scientists call this 'frequency-dependent selection' and it seems to be the mechanism that maintains other 'stable polymorphisms' such as that in Paradise Flycatchers, skuas and snails. The predators are better at spotting the commonest forms of snails, for example.

But no, for Ruffs it turns out this won't work. As ever, this idiosyncratic species is different in a critical way. The success of both the faeder and satellite mutations is strictly limited by the fact that birds that inherit two copies of the mutations do not exist at all. They cannot be born. So the mutations are continually being purged from the population. Yet somehow the mutations persist so it follows that the birds with one copy must be at some advantage, leaving behind more offspring. The Uppsala team calculates that these birds must on average produce 5 per cent more young. Where is that 5 per cent advantage coming from? Although females are apparently somewhat drawn to satellites on leks, and faeders can sneak in to mate, there is no good data to suggest that females mate preferentially with faeders and satellites rather than independents, so it's hard to see a 'sexy-son' benefit to counter the lethality in birds inheriting the mutations from both parents.

Nor does it appear that the birds with the inversion supergene live longer or survive migration better, though there's not much data either way, so a 'good gene' fitness benefit looks unlikely as well. A faeder male would have to survive three times as long as an independent to make up for its genetic disadvantages. It does not look like either mutation improves survival or reproduction. And there's another wrinkle. Given that the inversion supergenes are on a normal chromosome, females have them too. Each female will have one of three combinations of the gene: independent/independent, independent/satellite or independent/faeder. If a female has a copy of the faeder mutation, she tends to be smaller, lay smaller eggs and see fewer of her chicks survive. So the puzzle of how these mutations persist gets worse.

These 'inversion females' themselves face another risk. If they mate with 'inversion males' – that is to say a satellite or a faeder male – then a quarter of their eggs will fail to hatch. So it would make good sense for an inversion female to avoid such males like the plague. Lank spent four years trying to detect such an aversion among the inversion females and failed. In his aviaries, when given the choice of two males, the females were just as likely to mate with a satellite or faeder male as with an independent. 'I am massively disappointed as an adaptationist,' he tells me over Zoom, adding: 'The phenotypic gambit fails.'

However, there is a possible solution to this enigma. A clue lies in the gigantic size of the faeder males' testes. Recall that all Ruffs have big gonads for their size, compared with other wading birds. Testis size turns out to be the best answer to the question of how the inversion supergene survives. In relation to body size, faeders and satellites have huge testes compared with independents. This suggests that the advantage for these mutations lies not in getting more matings, but in the sperm competing to fertilise the egg inside the female after mating. As Küpper puts it, they invest in testes instead of territories.

Remember that unlike Black Grouse, female Ruffs, known as Reeves, usually mate with more than one male. Females tend to store sperm from these males for a few days in a special repository and then release them to fertilise the eggs when ready to lay. A female that mates with an independent and with a satellite a few days later (or vice versa) may find that the satellite's more voluminous ejaculation gives its sperm a better chance of reaching the egg when she is ready to unleash the stored sperm to do their job. Likewise for a faeder. If this is the case then satellites and faeders could mate less often than independents but still father more chicks.

So far I have not passed judgement on the Ruff's mutations, just tried to explain their oddities. But one of the greatest insights of evolutionary biology in the 1970s, associated especially with the biologist Richard Dawkins, was (paraphrasing JFK) to ask not what a gene can do for an organism, but what an organism can do for a gene. Bodies are genes' means of making more genes even more than vice versa. There are genes out there that have ways of getting themselves reproduced efficiently while harming the body's interests. Selfish DNA, they are called. If you take this gene's eye view of the Ruff inversion on chromosome 11, you might conclude that instead of seeking a way in which Ruffs benefit from having these mutations, we should be seeking ways in which these mutations thrive despite conferring no benefit. Clemens Küpper and his colleague Lina Giraldo-Deck eventually concluded that the inversion mutations have 'the hallmarks of a sort of a parasitic genetic element that is unable to persist without the ancestral arrangement'. They survive in a population consisting mainly of independents and cannot survive on their own, just like a behavioural tapeworm. The Ruff species would be better off without them. Lank remains doubtful, thinking that the benefit to satellites of being allowed to co-display with independents may be sufficient to explain the mutation's persistence.

'Females change mates within or between clutches'

The sun has come out and a few of the Ruffs are now lekking on the crowberry ridge, bowing, standing tall, fluttering in the air and fighting, their ruffs spread and their ear tufts erect. As I watch I ponder again the size of their testes – hidden away inside their bodies as in most birds. That Ruffs have unusually large ones was first noted by Emmanuel Baillon in the late 1700s. Writing to Georges, Comte de Buffon, he says: 'I know of no bird in which the appetite of love is more ardent; none whose testicles are so large in proportion.' In 1997 Tim Birkhead, the Sheffield professor who has most exhaustively explored the relationship between testis size and infidelity in birds, acquired a male Ruff to dissect and was similarly amazed. It told him immediately that female Ruffs mate with more than one male, as indeed they do. It is a general rule that species in which females mate promiscuously before each birth, like Chimpanzees, have larger testicles than species in which females mate with only one male in a season, like Gorillas. Chimps' testes are more than four times the size of Gorillas', even though male Gorillas weigh roughly four times as much as male Chimpanzees. That male Gorillas control harems of half a dozen females calls for only tiny testes (smaller than human ones) because those females mate only with him so his sperm have no competition. In sharp contrast, in Chimpanzee troops, females actively seek out matings with more than one male when in season – apparently to blur the issue of paternity and hence prevent the evolution of a habit of infanticide by incoming alpha males, which is a huge horror in the life of female Gorillas. Gorillas compete body-to-body; Chimpanzees compete partly ejaculate-to-ejaculate.

Fun fact: Ruff sperm have the longest tails of any sandpiper's sperm, another indication of fierce sperm competition in this species. In birds, large testes correlate with polyandry, with females mating with more than one male. In most avian cases, the purpose behind the female's choosing to mate with more than one male seems to be to try to get the best of both worlds. In the 1990s biologists discovered from DNA fingerprinting that lots of birds that were socially monogamous were anything but sexually faithful. Females formed pair bonds with faithful, devoted males occupying good territories but frequently sneaked off to get some better genes from the male next door (marry the man with a nice house but have an affair with the tennis coach?). In species where the male was needed to feed the babies, this enabled females to ensure the best possible genes for her offspring without losing the opportunity to get a male helping with the rearing of those offspring. And again it turns out that large male gonads are a good predictor of such strategies. Birds that lek, such as Black Grouse and Peafowl, do not therefore need big testes, despite polygyny, because females usually mate with only one male each: they do not use fathers as care givers at all.

Once again, Ruffs prove to be the odd bird out. They have huge testes and they are indeed polyandrous, with females mating with several males. Dov Lank, Terry Burke and their colleagues watched the Ruffs of the Oulu marshes in Finland over several springs in the late 1980s and managed to catch many of the birds and attach colour rings to their legs. They regularly saw individually identified Reeves mating with more than one male during a single visit to a lek, and mating with more than one male on successive visits to leks, enough to conclude that 'the majority of females change mates within or between clutches'. Half of the clutches of eggs they tested had more than one father.

Why? The father takes no part in rearing the chicks, so it cannot be a case of females choosing one male because he will make a good father and another because he will provide good genes. There is no need to pick a second best male as one mate merely because he's available for parental duties in this species. Nor is the female polyandry random. The Oulu study found that when females mated with more than one

male they chose both morphs – independents and satellites – significantly more often than by chance. It seems as if the females are
deliberately trying to have as diverse a family of chicks as possible –
which is why they are more likely than chance to choose a satellite for
the second mate if they have chosen an independent for the first mate.
Makes sense, but why for Ruffs and not for Black Grouse, I ponder as
I watch the Ruffs on this Arctic shore. There's an answer but I have not
found it.

The inversion mutations and the testes have deflected me from my
mission, which is to understand the infinite individual variety of Ruff
plumage. Remember that Ruffs do not just come in three colours. They
come in endless colours. The independents in particular can have any
combination of colours and patterns. Even the satellites are not all
perfectly white. Only the faeders lack this individuality. Lank has

Like Narcissus falling in love with his own reflection.

shown with his captive bred population that this individual plumage variety is heritable to at least some extent. But the precise genetic mechanism remains obscure.

Within the inverted supergene on chromosome 11 lies a gene called MC1R, which encodes a protein called the melanocortin 1 receptor. Mutations that mess up this gene entirely, known as missense mutations, have already been found to produce pale versions of birds in six other species, including the Red-footed Booby, the Arctic Skua and the Bananaquit. So the MC1R gene looks like a good candidate for explaining the diversity of colours in Ruffs. Intriguingly, the independents, satellites and faeders have slightly different forms of MC1R. Several other genes within the inverted region turn out to alter the way sex hormones are produced or the way the bird's body responds to them. It looks like MC1R interacts with other genes to generate the diversity of colours and patterns on both the back and the ruff. But a complete explanation for how Ruffs come to have such individual colours is still lacking.

11.

An Aesthetic Sense

And thence through seas and mountains and swift streams,
Through leafy homes of birds and greening plains,
Kindling the lure of love in every breast,
Thou bringest the eternal generations forth,
Kind after kind

LUCRETIUS, *De Rerum Natura*
Translated by WILLIAM ELLERY LEONARD

November, New Guinea

Seven months after the Black Grouse lekking season, I find myself at a
very different lek. I am a short distance south of the equator, in a dense
tropical forest on the side of a mountain in Papua New Guinea. My
soft-spoken guide, Logan, has brought me up a barely discernible path
through the trees to a special spot that only he knows, and we are
sitting on a log staring up into the forest canopy. The leaves in front of
our feet are littered with long, soft, flame-red feathers, evidence that we
are at a lek. But it is also evidence that the moulting season has begun,
so the peak of lekking is past for this year. Yet Logan is confident that
the males will still show up at some point.

The lekking bird we are waiting for is called the Raggiana Bird of Paradise, or 'kumul' in the Tok Pisin language. It is the national bird of this country and its image is everywhere in Port Moresby, from large sculptures in the middle of roundabouts to beer bottle labels to postage stamps. Today I have driven up into the foothills of the mountains in the hope of seeing the bird. Logan knows of the only lekking spot in this forest. For two hours nothing much happens as we sit quietly on the log. The cicadas buzz. A large swallowtail butterfly flits past. Orioles and parrots bugle and squawk. A loud growling sound comes from a Riflebird, itself a spectacular bird of paradise that uses its wings to make a sort of arch over the female – but it's a shy bird that's easier heard than seen. Then suddenly like a bolt from the red, a male Raggiana passes through the trees, calling loudly, but does not linger. Logan grabs my camera to snap a shot because he has a clearer view. After waiting a little longer we return down the path clutching some red plumes which can be used to adorn tribal headdresses that come out for celebratory sing-songs. Back at the car, our driver, Steven, in a modern version of the same tradition, sticks a couple of the plumes in the corners of his dark glasses.

Luck is on our side. When we return late in the afternoon, the lek is busy. Four males, each the size of a small crow, are flitting from tree to tree with loud 'Waa waa waa' calls, interspersed with softer buzzing sounds. They are showing off to a single female – or is it a juvenile male? It's hard to see what's going on through the leaves. My experience echoes Alfred Russel Wallace's account of when in 1857 in the Aru Islands off the west coast of New Guinea he became the first western naturalist to observe birds of paradise dancing: 'One day I got under a tree where a number of the Great Paradise birds were assembled, but they were high up in the thickest of the foliage, and flying and jumping about so continually that I could get no good view of them.' This was the Greater Bird of Paradise, a slightly larger species than the Raggiana and found further west. It is adorned with golden rather than red flank plumes, but the two species are close cousins.

The forty species of birds of paradise get their name from Portuguese and Dutch traders in the Moluccas in the 1500s who bought skinned specimens from local Malay traders who called them 'God's birds'. The reason for this, Jan Huyghen van Linschoten wrote in 1596, was that since the specimens had neither feet nor wings and nobody had seen these birds alive, then they must 'flie, as it is said alwaies into the Sunne, and keepe themselves continually in the ayre, without lighting on the earth, for they have neither feet nor wings, but onely head and body, and the most part tayle'. The truth was more prosaic: the bird skins came through secret trading links from the distant and unexplored shores of New Guinea and there the locals shot them with blunted arrows and removed the wings and feet before skinning the bodies to make ceremonial headdresses.

Today, however, rather than their dead skins it is their live dances, photographed and filmed especially by the indefatigable Tim Laman for National Geographic and the BBC, that give us glimpses of paradise. In 2010, close to where Wallace first saw the lekking Greater Birds of Paradise, Laman succeeded in setting up a leaf-concealed camera on the very branch where the birds displayed, which he triggered from a hide in a different tree, capturing the image he had long dreamed of: a Greater Bird of Paradise in full display against the dawn light with the rainforest stretching out below and behind. As an ensemble of avian and natural beauty it takes some beating.

The birds of paradise are sexual selection's masterpieces, the Monets and Matisses of female choice. They present a rich catalogue of bold colours, surreal shapes, eccentric accessories and wild dances that takes ornamentation for its own sake to bizarre extremes. Wallace referred to 'extraordinary developments of plumage, which are unequalled in any other family of birds'. And he had never seen most of them in action. He and his collecting companion Charles Allen found the people of

New Guinea reluctant to the point of violence to take him into the mountains and see where these valuable bird skins came from.

I strain my eyes trying to see the birds above me in the trees. At length I get a good view of an adult male, a gorgeous creature with a steel-grey beak, butter-yellow hood, bottle-green face, jet-black throat, fox-red body and two huge tufts of soft, long flame-red feathers on its flanks, which are at this moment being expanded inside its open wings and over its back into a plunging fountain of colour: backlit by the sun, the plumes and flapping wings give the impression that the bird is on fire.

The Raggiana species is one of the loveliest and largest birds of paradise but also one of the least weird. While displaying, unlike the shape-shifting Lophorinas, sicklebills and Parotias, it does at least have a recognisable head, tail and wings, albeit with curled wire-like feathers trailing from the tail when seen up close. It is a true lekking species, which many of the other bird of paradise species are not: they mostly display at solitary sites or 'exploded leks' where the birds are in earshot but not eyesight of each other.

I am here hoping to compare at first hand the Raggiana's lekking behaviour with that of the Black Grouse nine thousand miles to the north-west in a climate that is twenty degrees cooler and a habitat that is a lot less three-dimensional. Yet despite the vast distance and difference, once again, as with the Great Snipe and Ruff leks, I am struck by the similarities first. This Raggiana lek is in the same place every year, Logan confirms, just like the Black Grouse lek. It is not a noticeably unusual spot – a clump of typical, fifty-foot trees part way along a ridge. This lek attracts eight or so males on a normal day – similar numbers to Black Grouse leks. The action here is fastest at dawn just as it is in the north. Females visit the lek when ready to breed and are very choosy about who they mate with. There is not a lot of fighting, so it is clear that the purpose of display is to impress the female. (Why did Wallace seem to forget this?) And so on.

My first glimpse of a Raggiana Bird of Paradise.

What is so strange to me is that no other bird in this forest leks. There are parrots and loris and orioles and manucodes and Riflebirds and many other species, subsisting similarly on fruit, yet they see no need to gather for display or to evolve exaggerated plumage. Are they somehow less plagued by parasites? Seems unlikely. They are colourful, yes, but not crazy. Sexual display is a fairly brief episode in their year, whereas the Raggianas I am watching will, like the Black Grouse, dance every day for months on end. Lekking is a habit that females impose on males somewhat randomly, I think.

In California, San Diego Zoo has tried for years to breed Raggianas in captivity but found them to be reluctant breeders. Only when they tried mimicking a lek did the birds start to reproduce freely. Rather than housing the birds as pairs, they housed several males together and allowed them to form a lek, then introduced a single female to the

The Riflebird makes a growling call.

males' cage, removing her after she mated so that she could nest on her own. It turned out that it was critical to allow females a choice of mates, vindicating Darwin in the most direct manner. It would be great to know more about how lekking birds of paradise go about their business. But nobody has yet studied them in the wild in sufficient detail to record whether, like Black Grouse, the females mate with the same male, and only one; or whether, like Ruffs, they mate with more than one; whether they revisit the same lek year after year, how many seasons a top bird holds his position; and so on. There is no way that I can obtain any data from the distant and confusing glimpse I get of the lek in the trees above my head. But I am glad to have glimpsed it.

From here I had planned to travel into the Western Highlands of Papua New Guinea, where a dozen other species of birds of paradise live, right up to the tree limit in the chilly cloud forests. I had wanted to get there for another reason, to see the central valleys of the island, whose very existence was unsuspected until 1933, let alone the fact that they were inhabited by farming tribes that had never had contact with outside people from either the west or Indochina. But a series of travel complications – an earthquake damaging a runway and a broken down aircraft – prevented me getting there.

I had wanted to glimpse the extraordinary diversity of these birds by setting eyes on the Blue, the Superb, the Magnificent, the King of Saxony and other species. For if there is one lesson the birds of paradise reinforce more than any other it is the arbitrary nature of sexual selection. Or, to put it another way, the incredible creative imagination of the artist formerly known as female choice. After decorating the tail coverts of Peacocks, the wings of Argus Pheasants, the necks of Ruffs, the rumps of Sage Grouse, the tails of Black Grouse and the heads of Cocks-of-the-Rock, where the heck did this process get the idea of exaggerating flank plumes on Raggiana? Or enabling the Black Sicklebill to turn itself into a sort of enormous, shiny black flatworm pulsating atop a broken tree stump? Or turning feathers into antenna-like head wires as on Lawes' Parotia for part of its complex dance routine? Or making curved tail wires with emerald discs at the end for the King Bird of Paradise? Or sticking long ribbons of small blue flags on the side of the head of the King of Saxony Bird of Paradise? Or painting a bare-skinned skull bright blue on Wilson's Bird of Paradise? Or sticking long white feathers on the wrist of the wing in Wallace's Standard-Wing? It's almost as if evolution was on acid when it designed these birds, or as if they are consciously striving to be different in every possible way from each other.

At Yale in the Peabody Museum, some months before, Richard Prum had pulled out a drawer to show me the 'super-black' feathers sported by the Superb Lophorina, whose microscopic structural details include deep, curved cavities designed to capture all light and release none, like black holes. Such feathers, he and his colleagues found, can absorb up to a hundred times as much light as those on a crow or a blackbird, capturing 99.95 per cent of incoming rays when seen from head on. This rivals the super-black paint invented in 2014 known as Vantablack. The super-blackness of the feathers makes them impossible

Superb and Vogelkop Lophorinas. Photos by Tim Laman

to focus on for the human eye: they appear somehow blurred or insubstantial because there is no reflected light for the visual system to use. As a result, in the Superb Lophorina, the entire male bird seems to vanish, when it displays, into a sort of super-black oval hole in space-time, formed by its spread wings. Across this black void a fluorescent blue-green smiley face is projected with two dazzling dots above a wide lozenge formed by a breast shield of special, iridescent feathers and two bright eyelids. Thus shape-shifted, the bird bounces back and forth on a special log in front of a female with a series of loud clicks. Because of the ultra-black featureless feather background, it is almost impossible for the human eye to figure out what is going on, where the bird's head is hidden or how the blue-green smile appears suspended in nothingness. That females have generated something truly bizarre, through their aesthetic preference over many generations, is the best explanation of how this came about – yet another example of how much more creative, if not crazy, sexual selection is than natural selection.

In 2016 Ed Scholes and Tim Laman from Cornell University tracked down a population of what they thought was the Superb Lophorina with a distinctive song, on the peninsula that is shaped like a bird's head (*Vogelkop* in Dutch) in the extreme west of the island of New Guinea. They found big differences in the appearance and display. The Vogelkop Bird of Paradise – for it is now recognised as a new species – has a crescent-shaped black disc when displaying, giving it a frowny-face appearance as it dances, which it does by running sideways back and forth in front of the female, instead of bouncing. Probably the common ancestor of the two species had an intermediate form of display, which got exaggerated in two slightly different directions once ancient climate change separated the two species. Again, a tribute to the power of sexual selection to differentiate species.

In 2014 Bodo Wilts at Cambridge University and his colleagues examined the physical properties of the feathers on Lawes' Parotia, another shape-shifting bird of paradise with an iridescent neck patch.

The patch is so exquisitely designed with alternating layers of melanin rodlets and keratin that it can reflect light back only at a certain angle and that strongly, so it appears dull if seen from any angle but straight on, giving the distinct impression if you watch a Parotia dance of a light being switched on within the plumage.

One of the most striking things about birds of paradise – and Black Grouse and most lekking birds – is the fact that every part of the plumage is smart. The Parotias go through eight or nine different dances emphasising different parts of their plumage. These are not dull birds with one bright feature; they are decorated in different ways all over the body. Nor on the whole do they show a trade-off between looking smart and sounding smart or dancing smartly. In fact, a recent study of the behaviour and appearance of all the birds of paradise, using thousands of videos of them, found the opposite: the more complex the plumage, the more complex the song as well. Those that display high in the canopy in leks, like Raggiana, are both more colourful and noisier, whereas those that display on the forest floor, like Lophorina, have the best dance moves. But even allowing for this effect, the birds with the most complicated plumage also tend to have the most varied calls and the most intricate dance moves. This is an unexpected finding but provides the beginning of a theory as to how sexual selection accelerates evolution in birds of paradise. It creates redundancy. If a mutant male shows a small difference in only one aspect of appearance, call or dance, it may yet persuade a female to mate, so mutations can spread. As the Cornell evolutionary ecologist Russell Ligon and his colleagues write: 'If female birds-of-paradise make mate choice decisions based on sensory input from the multiple signals that comprise a composite courtship phenotype and information from those channels is correlated, then novel mutations changing the structure or form of a given ornament may occur without "necessary" information being lost.' The more the species evolves ornaments and dances, the more likely it is to go on evolving more of them till everything about the male is recruited to the cause of seduction. Instinctively, I feel that this argues against the good-genes theory of female choice and in favour of the runaway theory instead – to the extent that they are rivals at all. But I might be wrong.

'They seem to dissolve the barrier between art and science'

A few days later, a few degrees further south, still on the old Gondwanan continent but this time south of the Torres Strait in northern Queensland, I am standing in dense tropical rainforest, a short distance from a track, listening to a bird pour forth its heart in an almost deafening series of whistles, squeaks, squawks, buzzes and clicks, many of them perfect imitations of other forest birds. It is hard to see the songster but eventually I get a good view of a plump khaki-green bird with a stout beak that ends in a small hook, from which it gets its name: the Tooth-billed Bowerbird. It looks similar to its cousin the Spotted Catbird, one of the birds it is currently imitating. But it is not the song of the Tooth-billed Bowerbird that has brought me here, rich though it is, nor the fact that it operates a sort of exploded lek, in which males are clustered just close enough together to hear each other when singing. No, what I am here to witness is what lies below the bird, on the floor of the forest. Immediately beneath its song perch, the male has cleared a patch of the forest floor of all sticks and dead leaves and then neatly placed an array of fresh green leaves, all turned upside down to show their silvery undersides. This bird is an artist, albeit of a very simplistic kind. To persuade a female to bear his offspring, he must sing his heart out, yes, but he must also make an installation. He collects leaves and displays them neatly on a clean forest floor, regularly refreshing them. Sexual selection has, in this species, stepped outside the body and altered the way the bird behaves with respect to objects around it. That is a key moment in evolution.

The Tooth-billed Bowerbird, which spends much of the year eating leaves, probably retains many of the features of the first ancestral species of catbird to start decorating its display 'court' in this way. Other species of bowerbird have since gone much further. In another dense rainforest an hour's drive away, and a mile's hike up a rough track from the nearest dirt road, I find myself sitting at the base of a tree, looking intensely at a small, dead branch. The air is warm but the shade prevents it being oppressively hot. Ants and mosquitoes are trying to spoil my fun but not very hard. James Boettcher, my guide, has brought me to this spot, to watch this particular perch, because he knows its significance to an extraordinary bird. The branch is about a

The Tooth-billed Bowerbird spreads leaves on the bare forest floor.

foot off the ground, and it links two huge pyramids of sticks, each about three feet high and each piled up into a cross between a maypole and a ziggurat. There must be tens of thousands of twigs in each and they were all clearly put there by somebody or something. At either end of the bridging branch that joins the two piles, there is a neat heap of soft, grey lichen fragments, topped with a small garnish of white flowers. Whoever built this structure wants it to be beautiful.

Suddenly James nudges my elbow and points. The artist is coming. A delicate, fairly small bird, olive green on the back but bright, golden yellow on the front, and with a long, yellow, slightly forked tail, lands on the perch. In its beak is a small, white flower that it places neatly on one of the two heaps of flowers and lichens before appearing to stand back and admire its beakiwork. It flies up into a tree, emits a weird, mechanical buzzing sound like a clockwork toy – which is soon imitated and echoed by a Tooth-billed Bowerbird nearby – before flying off to fetch more lichen and flowers. This is my first glimpse of the Golden Bowerbird and although I knew what to expect I am still left incredulous that such a small bird should have built such a massive structure. We wait another two hours and get two more glimpses of this elegant and colourful little architect-cum-flower-arranger at its bower. I know I am very lucky to see the bird. A mostly silent species that lives deep in the tropical rainforest, its bowers are known only to a few guides, who guard them carefully. A few years ago, near here, three adult males disappeared in a short space of time, soon after a client hired three different guides to show him the birds using a differ-ent name each time. The guides compared notes and worked out what happened: it seems the 'client' was a bird thief, probably well paid to catch and cage birds for collectors in Arabia. I promise to tell nobody where this particular bird lives.

In every bowerbird species, it is the males that build bowers, and the females that visit and judge them before breeding with no help from the male. The bower is no nest; it is a seduction device of no other practical use. It is playing exactly the same role as the plumage of the Black Grouse does – charming the female into mating. Something about inviting her into a flower-decked tunnel of love between two huge piles of sticks will tip the balance in this male's favour, though not today, as no female shows up. What really interests me is that the bower and the collection of flowers are evidence that sexual selection has jumped the shark in this family and divorced itself from bodily adornment into artefact adornment. Although they sing and dance as well, most bowerbirds try to impress females with things they have collected or built, not things they have grown. Sexual selection has begun to operate outside the bird's body, discriminating between males not just on how they look and move but on what they have constructed and gathered. Bowerbirds are not the only creatures that use art and artefacts for seduction; so do large, bipedal, African-origin apes: you and me. Hence my fascination.

As long as I can remember it has been a dream to see these bowerbirds and their bowers. Eccentrically, I think they are among the most thrilling sights any naturalist can witness, because they seem to dissolve the barrier between art and science, as well as between people and birds. The eighteen species of true bowerbirds live only in Australia and New Guinea and some are only just beginning to be known in any detail, living in remote mountain regions of New Guinea. Each species builds a different structure and indeed some isolated populations build their own peculiar bowers unlike those of their conspecifics in the next province. Bowers range from the extremely simple – the Tooth-billed Bowerbird's plain, leaf-strewn 'court' – to the massive and complex. The Golden Bowerbird's paired maypoles can be six feet high, and the Vogelkop Bowerbird in western New Guinea builds a truly extraordinary thatch-roofed hut with an open doorway, in front of which it places neat piles of fruits, flowers and plastic items carefully arranged in separate piles by colour. The first westerner to see these intricate bowers in 1872, Otto Beccari, refused to believe his Papuan companions when they told him the structures were made by birds and not by people.

The male McGregor's Bowerbird, in New Guinea, takes years to build a truly enormous maypole, in the centre of a perfectly circular moss-lined arena, a metre across, with a raised moss rim, then decorates the maypole with baubles made of flowers and fruits, like a Christmas tree. It is probably the largest construction by a bird. His display routine is equally intricate and complex, beginning with mimicked sounds – which include laughing children and barking dogs as well as other birds – before moving on to a sort of hide-and-seek game with the female around the maypole, culminating in the sudden and colourful explosion of his yellow crest, accompanied by a series of crazy movements. Truly, the female McGregors must be having an evolutionary laugh as they contemplate the ordeal they have imposed on the males of their species.

'Why did perspective evolve in bowerbirds before it did in humans?'

Another day, another species. I am standing by a path at the edge of a mountain rainforest hundreds of miles further south, on the border of Queensland and New South Wales. But in truth I am also just a few yards from the car park of a hotel spa. It is just after dawn and a soft mist presages a warm day. Beneath some bushes, between a clump of grass and some dead ferns, is a structure that is much smaller than the Golden Bowerbird's bower, but more precisely shaped. Grass stems have been fashioned into a U-shaped tunnel-like avenue, almost closed at the top and open at each end. In front of it lies a broad platform of grass stems – known as the court – across which are scattered bluish flowers, a few blue parrot tail feathers and about twenty bright blue man-made things: a plastic bottle top, the cap of a ballpoint pen, several torn fragments of 'Mentos' mint-sweet wrappers and something vaguely medical. The owner and creator of this bower is sitting in the tree above, calling harshly like a jay: a Satin Bowerbird. A chunky, dark bird larger than the Golden or the Tooth-billed Bowerbirds, it is the size of a small crow – to which bowerbirds are distantly related – with a stout whitish beak and a short tail. It looks black from a distance but closer up it is a deep, shiny, indigo-blue. The iris of its eye is bright violet.

As I watch, partly concealed by a thorny bush, he flies down silently and hops over to his bower, bringing a couple of extra grass stems in his beak to add to the construction. He pops into the bower, does a little housework, painting the sides with berry juice, then picks up some unwanted dead leaves and flies off. An hour later he is back down at the bower with a mouthful of whitish flowers. He puts them down, stands back to take a look, changes his mind and rearranges them. He moves some of the blue plastic things about, then goes and grabs some more grass stems and slots them into the walls of the bower.

After this he returns to his high perch and calls again, launching into a perfect imitation of a King Parrot when one of those flies by. Then, suddenly, he stops in the middle of a burst of calling and flies straight down to the ground next to his bower. A slightly larger, green-backed, mottled-fronted bird appears through the bushes: a female. The male hops over to his art collection and picks up a bottle top in his beak, unleashing a continuous, conversational stream of buzzes and whistles, a bit like an old-fashioned, dial-up modem making a connection. She steps into his bower, confined by its tight sides. His excitement mounts. But when he moves too close she jumps out and flies off.

Later in the day I watch a different species of bowerbird, the Regent. Though it is the breeding season I don't find a bower and the locals tell me they rarely see them. Regent Bowerbirds construct flimsier bowers than Satin Bowerbirds, decorate them much less and defend them for just a few weeks. Almost as if the bower matters less to this species, and indeed the male displays to the female away from the bower before leading her to it. Here's the intriguing fact: male Regent Bowerbirds are stunningly beautiful birds, boldly patterned in black and the brightest possible yellow, turning to orange on the top of the head. The bird looks like it has electric lighting in its plumage. Does the Regent – and its three close relatives from New Guinea, which all have slightly different patterns of the same black, orange and yellow plumage – still retain more of a tendency to do its sexual selection on plumage rather than bower quality? One of them, the Flame Bowerbird, is known as the most brightly coloured bird in the world. In general, the bowerbird species with the most elaborate bowers have the least spectacular male plumage and vice versa. Within the family, a tendency to court females through art has led to a reduced tendency to court them through feathers. Moreover, an analysis of the brains of bowerbirds has found that the ones with the most complicated bowers have the largest cerebellums – this being the part of the brain responsible for procedural learning and motor planning.

It is on my last day in Queensland that I see the fullest sequence of bowerbird behaviour. This time, James and I are on the outskirts of a small town, set in farmland, on the side of a busy road. On one side is a mango plantation; on the other, a cemetery. Under a scruffy, small tree at the edge of the cemetery, right by the roadside and a few yards from the nearest grave, is the most bizarre art collection at a bower I have yet seen. The bower is a tunnel made of small sticks, much like that constructed by the Satin Bowerbird but bigger and more enclosed overhead. At either end, and to the sides, on the 'court', lies a truly massive spread of coloured objects. Anything white seems to have caught the collector's eye: from snail shells to white stones to bits of plastic. Transparent is good, too, so there's heaps of broken glass and bits of see-through plastic. Grey is not bad, so the plastic toys of a nearby war-like child have been brought here: half a tank, a couple of artillery pieces, a small plastic hand grenade. (Bizarrely, my 2019 textbook on bowerbirds records a similar hand grenade being found at a bower in the next town a few miles away, so perhaps it was the same toy.) The little boy who lost these items probably has a sister, because the pièce de résistance at the front of the bower is a magnificent plastic tiara glistening with fake diamonds. She has probably been scolded by her parents for losing it. Then to add a bit of colour, the bird has brought along some green berries and fruits and a few bits of bright red plastic, plus a healthy helping of ripe, red chilli peppers: real ones. The red items are mostly placed to the sides of the bower rather than at the ends. The green ones are between the red ones and the white ones. The arrangement is definitely non-random.

James unfolds a small camo-pattern hide so we can watch the owner of the bower at work, and we climb in. We pray that no mourners turn up for a funeral and start to question why two grown men are crammed into a camouflaged tent within the grounds of the cemetery. Within minutes the bird has flown back. It is a Great Bowerbird, which despite its name is no bigger than a Satin but has a longer tail and quite different plumage. It is pale grey-brown all over with a spotted back and wings – nothing much to look at, although later we are lucky enough to glimpse its bright, lilac-coloured headdress, which usually lies hidden under grey feathers on the back of the head. The male hops about, picking up items and rearranging them. He gives the tiara a close and proud look. Then suddenly he starts buzzing with excitement – literally, in the sense that he emits similar 1990s-modem noises to those made by the Satin Bowerbird. A female has arrived.

For the next twenty minutes James and I are treated to a virtuoso performance. The female repeatedly steps inside the bower and seems to check its features, occasionally rearranging a few sticks, while the male hops about outside, bowing, raising his tail, running, hopping, jumping and buzzing. At one point he picks up a chilli pepper and proudly parades it in front of her while she stands inside the bower.

There is a rather intriguing insight from science of relevance here. Deakin University's John Endler in 2010 made a fascinating discovery

A toy hand grenade and a tiara collected by a Great Bowerbird.

about Great Bowerbirds. Working in a wilderness area where litter was not available, he and his colleagues noted that the birds arranged their white and grey stones, skulls and shells, on their courts at the ends of the bower, in such a way that the larger objects were further from the bower and smaller objects closer to the bower. This meant that, when viewed from inside the bower, the objects all appeared to be the same size. The effect is one of forced perspective, making the court appear more regular and perhaps smaller than it is and making the male on it seem correspondingly bigger. Endler did various experiments to show that this is a deliberate decision on the part of males. Males spend a lot of time alternately looking at the court from the inside of the bower and rearranging the objects, and when Endler and his colleagues rearranged courts so that larger objects were closer, and smaller objects further away, the males promptly changed them back again. Forced

perspective seems to matter to the birds, because males whose arrange-
ments were most carefully graded by size seemed to get more matings,
and males tended to wave objects (like chilli peppers in our case)
towards the females in a way that made them loom large against the
regular background. Endler asked, cheekily: 'Assuming that bower
building occurred in bowerbirds before the fifteenth century, why did
perspective evolve in bowerbirds before it did in humans?'

In our cemetery bower case, this means tiaras and hand grenades
should therefore be placed furthest from the bower – and indeed they
are. The female is within the bower looking out at the male, with his
chilli pepper, across the object-strewn court and the size gradient of the
objects is giving her a very regular background. Then suddenly the male
briefly reveals his lilac crest, turning his nape towards the female and
giving her a startling splash of colour against the monochrome court.
The excitement seems to grow. But just as we begin to hope we might
see a mating, she flies off. After a short while James and I head into town
for some breakfast, leaving the cemetery to its deceased inhabitants.

'The males took all these gifts from heaven straight to the bower'

Although from the start bowerbirds' bowers were clearly seen by biologists as something to do with mating (and by local New Guinea
tribesmen, who like to point out that they always understood the
importance of bride price), exactly what went on at bowers remained
largely unknown until Clifford and Dawn Frith began studying the
birds in exhaustive detail in Queensland in the 1970s, colour-ringing
individual birds for identification, and staking out the forest so as to
map their movements, then witnessing the series of intricate, progressive steps that each sex must take to get to the point of mating. In the
Satin Bowerbird, females visit and revisit several bowers over several

weeks both before and after they build a nest and become ready to conceive. After nest-building they narrow down their preference to just a few bowers then allow copulation at just one, after which they head off to hatch the eggs and raise the chicks on their own with no help from the father. Competition between the males is fierce, with individual males raiding each other's bowers for ornaments and vandalising rivals' bowers. The birds are very slow to mature: a male Satin Bowerbird looks like a female till five years old and then gradually gains the satin-blue plumage, so he may be eight or ten before he gets a chance to build and defend his own bower. Indeed, he usually has to inherit a site, with bowers persisting on the same site for decades, under several owners. If he tries to set up a bower within a short distance of an existing male's bower, it will be destroyed in short order.

Starting in 1999, in a neat, if slightly unkind, experiment, Seth Coleman, Gail Patricelli and Gerald Borgia manipulated the attractiveness of up to twenty-eight Satin Bowerbird bowers over two years at a site in New South Wales. Noting that females visited bowers and checked them out early in the season well before nesting and when males were not yet likely to display, the scientists decided to enhance the attractiveness of half the bowers and see if it affected later mate choice by females. In each case they paired an experimental bower with a control bower that had been similarly attractive in the previous year. In the first year, 1999, they then put twenty blue plastic tiles on the 'court' platforms of each of the experimental bowers. In the second year, 2000, they placed another twenty tiles and fifty strands of blue plastic close to the bower. The males took all these gifts from heaven straight to the bower within two hours. (Imagine the joy of seeing seventy new blue things nearby one morning, blue being normally a very rare colour …) To prevent stealing, the scientists then fixed each of the new decorations in the place where the male had placed it. Also, and this is the mildly unkind part, they checked the control bowers twice daily, confiscating any of the tiles or strands that turned up there.

(Imagine the frustration of adding blue objects only to have them vanish …) The scientists thus greatly enhanced the size of the blue-object art collection at half the bowers.

So now within each pair of bowers, one should be much more impressive to a female than the other, at least on her early-season visits when she sees no male display. This was indeed the case but because they had colour-ringed almost all the sixty or so females in the area, the scientists were able to detect an age effect: young females were much more impressed than older ones by the enhancement. Yearling females were much more likely to return to that bower to watch the male's display both before and after her nest-building. They were three times as likely to mate with an experimental male, with his enhanced art collection, as a control male in the first year of the experiment, while two-year-olds were somewhat more likely and three-year-old or more females were no more likely. In the second year, with the addition of blue strands as well as blue tiles and with slightly more pairs of males, the effect was even stronger. Yearlings were more than five times as likely to mate with an experimental male and two-year-old females almost four times as likely. Again, the oldest females were not noticeably impressed by the blue-litter enhancement. The experiment shows that female choice is probably a lot more complicated than it appears on the surface: with significant variation among females in how they express their preferences and an unexpected variation by age.

A male Satin Bowerbird who – like an eligible Mr Darcy – has a good estate, keeps his pile in good order, prevents its art collection being stolen and sings and dances with flair must be in with a chance of finding some wives. All this is consistent with the idea that bowers and displays are both – for the females – a means of sorting some good genes for your offspring, which is why they evolved. But could they also – or instead – be a runaway phenomenon? Namely, that females have to choose the most elaborate bowers or the biggest collection of blue objects because if they did not then they would risk having sons that got no mates. There might be no other rhyme or reason than this. Male bower-building and female choosiness could have co-evolved in an arms race and the bowerbirds are embarked by chance on a train that they cannot get off. Building two six-foot ziggurats of sticks in the middle of a rainforest, when you are the smallest species of bowerbird, strikes me as a pretty random fate. Collecting more blue plastic things than your rivals, if you are a Satin, or more toy tiaras and hand grenades, if you are a Great, might not be any good at showing you have good genes for evading predators or diseases; it might just show you are good at collecting blue things and tiaras. We're back to the big argument in sexual selection and frankly, this is the same old argument as in lekking grouse. Both effects are likely at work.

12.

How Mate Choice Shaped the Human Mind

Love looks not with the eyes, but with the mind.

WILLIAM SHAKESPEARE, *A Midsummer Night's Dream*

December, the Pennine hills

It is midwinter. I am watching a small flock of Black Grouse feeding on the buds of birch trees not far from the lek. It is probably the old team from the lek with some young recruits. Has Black Spot survived? And Black Bar? And Wonky Tail? Hard to tell while their tails are stashed away. I will find out next spring. (As this book is going to press, I can report that Wonky Tail did survive and started the lek the next spring in a central position. But he looked bedraggled, had much smaller red combs than his rivals and displayed less often.)

But I am thinking about human beings. These male birds carry most of their sexual ornaments all year and for life: blue neck, white bum, black and white wings. The imperatives of one fortnight in April have saddled them with conspicuous plumage forever. This is what Charles Darwin was thinking about when he emphasised sexual selection as a far more powerful force in evolution than anybody else realised: it can alter a species drastically, diverting its plumage and behaviour towards success in courtship; survival be damned. Can the same be true of me?

Wonky Tail photographed in 2021, 2022, 2023 and 2024.

Are there aspects of my body and my mind that are designed not for survival but for courtship, that are authored not by natural selection but by mate choice? Of course – but how far does it go?

This is not a book about people. I am interested in birds for their own sake, and it irritates me when – all too often – acquaintances assume the only thing that is interesting about the displays of birds is what they say about us. We are just a dull primate. If you want real beauty and extravagant ornament, and want to understand evolution, forget the human species and look elsewhere, to birds especially. But, having engaged in courtship myself, admittedly a long time ago, watched other human beings in various stages of courtship, especially on film, and talked to others who have studied the process up close, I feel I should speculate about *Homo sapiens* briefly.

Here is what is known about the human mating system. First, people don't lek. Males don't gather at traditional spots to display communally and ritually to females, let alone watch just one of them then mate (on site) with all the females. Only a few mammals do lek: Fallow Deer, Blackbuck antelope and White-eared Kob, a species of African antelope. Unlike Black Grouse, human beings are not a highly polygamous species with one male getting to mate with most of the local females, leaving most of the other males to die celibate. Nor do human males display on widely separated but fixed display arenas like Argus Pheasants. Nor do they grow exaggerated and bizarre appendages for use in courtship dances. They are dull, utilitarian, ugly creatures by the standards of birds. And they mostly form pairs.

But that does not mean they are untouched by sexual selection. Remember the lesson of the Crested Auklets in the Pacific Northwest: sexual selection happens in pair-bonded, monogamous species too, only it's mutual – males selecting the best plumaged female they can get and vice versa. Darwin considered this possibility briefly: 'It is again possible that the females may have selected the more beautiful males, these males having reciprocally selected the more beautiful females.' But he rejected it. Then for decades biologists pursued the frankly idiotic idea that the reason kingfishers or penguins sport bright colours in both sexes is so that they don't make a mistake and mate with the wrong species. Indeed to this day a lot of people think sexual selection is always about stark differences between the sexes. A distinguished paleontologist once told me that he had found no evidence of male and female dinosaurs being different in size, so sexual selection was irrelevant to dinosaurs. I sent him the paper about small, flying dinosaurs called Crested Auklets and how they are shaped by mutual mate choice.

Not that human beings are strictly monomorphic like auklets: scientists report that males are on average significantly larger than females, grow much more facial and body hair, have considerably greater upper-body strength and lack the swollen breasts that are so characteristic of adult females in this species. But compared with their closest relatives, the other great apes (Orangutan, Gorilla, Bonobo and Chimpanzee), these sex differences are comparatively modest. A silverback male Gorilla weighs twice as much as a female. And compared with them, human beings are much, much more inclined to form lengthy pair bonds. Gorillas and Orangutans form harems, while Chimpanzees and Bonobos live in multi-male, multi-female troops. Scientists have discovered, by contrast, that a male human being may spend years, even decades, mated to the same female. This is most unusual for a mammal; much more typical of birds.

However, just as DNA fingerprinting surprised scientists in the 1990s into the realisation that socially monogamous birds are not always sexually faithful to their mates, so the same appears to be true of human beings. Both sexes will in some cases furtively enjoy sexual liaisons with others if the opportunity arises, while still maintaining a pair bond to raise the offspring, and males will sometimes unwittingly rear offspring that they have not fathered – a habit they share with many socially monogamous birds such as Skylarks.

Moreover, as is also true among some birds, simultaneous polygamy of the Gorilla kind does occasionally occur in the human species, nearly always in the form of polygyny (one male mated to several females) rather than polyandry. It still often involves long-term bonds, though, not leks. Harem formation of this kind appears to have been rare when human beings were all hunter-gatherers, but became more common and more extreme when agriculture brought greater inequality of wealth, and then polygamy became almost the norm in the predominantly pastoral societies of central Asia, the Middle East and parts of Africa, where owning two hundred cattle or goats was not much harder work than owning twenty, and being the second wife of the two-hundred-cattle man was probably a better life for a woman than being the only wife of the twenty-cattle man. Indeed, polygamy survives as a legal option in some nations that are descended from these pastoral societies. In most modern nations, however, polygamy has become much rarer again: scientists are unsure if this is for economic reasons, relating to the costs of maintaining a home and rearing children, or because of preaching by moralists against polygamy and in favour of the rights of women (and twenty-cattle men), or because of legal proscription. Whatever the cause, pair bonds are again the norm in Europe, the Americas, and most of Asia and Africa.

The evolutionary cause of this is almost certainly that bonding appears to be a crucial part of child rearing in the human species, just as it is in Skylarks. Given how utterly helpless human babies are – little more than external foetuses for the first year – and given how long they take to learn to survive on their own – the best part of two decades – it has remained true throughout history that most cultures regard parenting as a two-person job. That is just as true today in a prosperous city, and with taxpayer-funded social services, as it was twenty thousand years ago in a Paleolithic foraging society on the savanna.

In other words, it appears that it is a human instinct to form a long-term pair bond to raise offspring. The existence of a strong instinct of sexual jealousy in both sexes is a sure indication of the importance of pair bonding in our species; it is the force that undermines all attempts at creating free-love communes, for example. Sharing either a wife or a husband with others just does not come easily to people – let alone to the spouses in question. In addition, human beings have a highly unusual habit – shared with Bonobos but not Chimpanzees – of having sex pretty well continuously once adult, even when not ovulating. This, and the fact that ovulation is undetectable to males, seems to point towards human beings having come to use 'recreational' sex as a means of reinforcing and maintaining pair bonds.

But I reiterate: just because human beings are usually a pair-bonding species does not mean that they are not subject to sexual selection. Picking a partner for life is a big decision, and one that it pays to get right if possible, so human beings are very picky. Studies show that while men may be more keen than women to have one-night stands with strangers, when it comes to selecting life partners both sexes are equally choosy. Male human beings may fantasise about short-term sexual liaisons (more than females) but they are also, like human females, obsessively selective about those they form long-term pair bonds with. Indeed, it's the dominant theme, and the happy ending, of most romantic fiction whether in print or on screen. Rom-com writers know their audience.

The evidence that many features of human beings have been selected by their mates has been staring us in the (bearded in one sex) face for centuries. The most striking and disturbing example is probably the human mind itself. Did it grow massively bigger in a comparatively short time not to help people survive but to help them seduce each other?

'Acquired traits had sunk into the hereditary substance'

Sexual selection may explain the start of art. As Darwin put it in the *Descent of Man*: 'The playing passages of bower-birds are tastefully ornamented with gaily-coloured objects; and this shews that they must receive some kind of pleasure from the sight of such things.' The Satin and Great Bowerbirds that I watched courting females with bottle tops and chilli peppers are of course driven mostly by instinct, says the conventional wisdom, while human beings have culture. Hmm. I think both ends of that claim are partly wrong: people are driven more by instinct than we admit; and bowerbirds have more culture than we assume. Bowerbirds are unusually large-brained birds. The Cambridge zoologist John Madden surveyed the behaviours of bowerbirds and concluded that 'despite a paucity of data in comparison with primate studies, it could be argued that bowerbirds may be considered to fulfil the same criteria on which we base our use of the term culture when applied to our close relatives, the great apes'. For example, a Spotted Bowerbird that found itself swept off course by a storm, ending up in Satin Bowerbird country, learned to collect blue items instead of its usual white, green and red.

As for whether people have instincts, plenty of experiments show that people have innate tendencies and that the way these work is often through making people more likely to learn some things than others, so culture and instinct are not opposites, but work together. Nature operates via nurture, not versus it. As the anthropologist Joe Henrich has documented, this means, for instance, that when people make mistakes they tend to be in an adaptive direction such as mistaking safe animals for dangerous ones rather than vice versa. Learning is a means of evolving.

This means it is highly likely that when smaller-brained human ancestors first began decorating their bodies, their homes, their clothes and their tools, it was likely very much more at the behest of instinct rather than the result of any kind of rational calculation. Bowerbirds might help us understand how that happened.

Take, for example, the Acheulean handaxe. This is the name given to a widespread tool used mainly by *Homo erectus*, which turns up again and again in the archaeological record, over an immensely wide area from Europe to southern Africa to Asia and over an unbelievably long period from almost 1.8 million years ago to just 300,000. It's teardrop-shaped, symmetrical and sharp-edged. Years ago I spotted that the one that sits on my desk – my wife found it on eBay, but it probably came from North Africa originally – was exactly the same size and shape as the computer mouse I used at that time. This sent a shiver down my spine: two objects designed to fit the human hand but separated by at least half a million years. Nobody knows quite what Acheulean handaxes were mainly used for but skinning animals, slicing vegetables and whittling wood are probably on the list, though throwing them as sharp-edged projectiles has also been suggested. It is not immediately obvious to us today why this design was so well suited to some particular task, let alone why it worked equally well in places as far apart as Africa, Europe and India.

The far greater problem the handaxe poses is why it never changed much either in space or time. It experienced more than one and a half million years of extreme technological stagnation. Every other technology hominins invented, from the spear thrower to the Swiss army knife has evolved pretty rapidly thanks to innovation. Not the Acheulean handaxe. That it could be so universal and change so little over such a long period – hundreds of thousands of generations – defies cultural explanation. If it was a product of culture it would surely show fairly rapid change and significant geographical variation: that's what culture does. Yet archaeologists have consistently and universally assumed that the Acheulean technology was indeed purely the product of culture: within-tribe imitation according to a social tradition. They have

My handaxe and my mouse.

assumed that because, well, they always assume that about human beings. The possibility that we are looking at a product of instinct is just never even considered. In the words of four scientists arguing for a different explanation: 'When nonhuman animals display complex behavior, the default assumption is that it is under genetic control. For complex behavior in humans and other hominins, however, the default position is to invoke culture and not genes.'

It was the anthropologist Rob Foley who first made the suggestion in the 1980s that Acheulean handaxes might be partly the product of genetically inherited habits rather than of purely cultural learning. He was largely ignored. Two evolutionary anthropologists, Pete Richerson and Rob Boyd, took up the idea in 2005, describing it as 'bewildering' that so little variation would appear in a cultural tradition. Then four anthropologists based in Canada and the Netherlands, led by Raymond Corbey of Leiden University, revisited the debate in 2016, adding several new strands of argument. They pointed out that scientists strug-gle to explain the sudden arrival of fast-changing and cumulative stone technology in the upper Paleolithic, after about three hundred thousand years ago, usually reaching for a 'cognitive upgrade or increased population size' as the trigger, neither of which is persuasive. Instead, Corbey argues, if the previous technology was largely instinctive then the arrival of cumulative cultural innovation is more easily explained: it was a switch from expressing an instinct for learning something to a pattern that was very much more dominated by social learning. And Corbey points out that evolution is fond of endowing even highly intelligent animals such as bowerbirds with genetically determined behaviours if only because it is a short cut for the individual to getting the behaviour down pat. When I contact Corbey to explore this further, he reminds me that 'the relevant data base is a gorgeous challenge. It comprises a few million known handaxes, and billions of unfound ones, on a conservative estimate.'

Of course, most cultural anthropologists reacted with disdain to this genetic theory, rabbiting on instead about 'preferred cultural conservatism' and 'the psychological bias for majority imitation that subsequently became a social norm', which is little more than a restatement of the problem. Corbey responded to his critics by pointing to the 'Baldwin effect', an idea proposed by the American psychologist James Mark Baldwin in 1896. Baldwin said that animals which adopted new habits or found themselves in new environments would then find that any genetic mutations which helped them cope with those circumstances would be favoured. As those mutations spread, the new habits would become more heritable and instinctive, therefore. So what starts as a fully learned behaviour could end up as a partly inherited one. This might superficially resemble the inheritance of acquired characteristics, but it would not be. In Baldwin's words: 'It would look as though the acquired traits had sunk into the hereditary substance in a Lamarckian fashion, but the process would really be neo-Darwinian.' Jean-Baptiste Lamarck famously proposed a theory of evolution in which offspring inherited traits that were acquired during the lifetime:

a body-builder would have muscular children, for example. But the Baldwin effect is subtly different. So, for example, by taking up the consumption of milk, human beings in a few parts of the world put themselves unwittingly in a good position to promote mutations that allowed the digestion of lactose in adulthood, and sure enough the genomes of people changed. Culture can therefore drive genetic change. I cannot emphasise enough just how much both most social scientists and most evolutionary biologists struggle to see this point.

In the case of handaxes, the significance of this insight is that if early hominins started knapping stone tools, any mutant human being with a genetic propensity to be good at this skill from the start would thrive, and leave more offspring, so the skill would itself gradually become more heritable. As Corbey puts it: 'If phenotypically plastic individuals grow up time and again, over hundreds if not thousands of genera-tions, in a technological niche while manipulating stone, and provided that the cost/benefit ratio is right: would not selection in the long run favor features of the organism befitting their technological capacities, so crucial for survival?'

This goes so against the grain for cultural anthropologists – and most of the rest of us – who think of human beings evolving away from instincts and towards culture, that they and we struggle to get the point. But evolutionary biologists who have looked into the mechanics are agreed that the Baldwin effect is a real phenomenon. So even if *Homo erectus*'s ancestors did not have a genetic tendency to enjoy making stone tools, their descendants probably did. Even if the first bowerbirds were simply copying successful rivals, their descendants found themselves with highly specific instincts to (learn to) build highly specific bowers. It is no more fanciful to suppose that *Homo erectus* had an instinct to make handaxes than that a Satin Bowerbird has an innate instinct to make a highly complex bower to a particular design. At the very least the hypothesis should be taken seriously. This does not mean handaxe-making required no learning; bowerbirds still have to learn how to perfect the building of bowers by watching others and practising. But it is directed learning, towards an instinctive, adaptive end. I repeat: a lot of learning is like that – nurture reinforcing nature, rather than overcoming it.

In 1998 Marek Kohn and Steve Mithen had a still more speculative thought about handaxes. The force that shaped that instinct – and the entire Acheulean technology – might have been sexual selection, not natural selection. This is the case with bowerbirds' bowers, clearly: they came into existence through sexual selection, not natural selection, by improving the chances of reproduction, not survival. Kohn and Mithen pointed out that handaxes are far more symmetrical than they need to be. They seem to have been deliberately made so, despite an asymmetrical tool being just as useful in butchery and a lot easier to make, probably even easier to hold. If female bowerbirds choose a mate based on how well he constructs a symmetrical bower, perhaps female hominins chose a mate based on how he constructed a symmetrical handaxe. Why not? It's as plausible as any other explanation. The handaxe may have been a work of art made by a male and intended to impress a female – or vice versa. To see human females judging a potential mate on his ability to shape a really good handaxe may seem absurd but it's no more so than a female bowerbird judging a potential mate on his ability to forge a bower from sticks and decorate it with coloured objects. Both are difficult tasks that could enable females to find good mates. Both generate products that have limited practical use. As Darwin kept emphasising, and being ignored for it, sexual selection is not a peculiarity in one small corner of the natural world: it's a massive force for evolutionary innovation that we keep overlooking. Perhaps the handaxe was only partially a practical tool, but also an ornament, a seduction device. Just like a bowerbird's bower. And perhaps Darwin was right to look to sexual rather than natural selection to explain much of human behaviour.

'We're all descended from artists because art was sexy and art was romantic'

Like bowerbirds, human beings use decorations and ornaments to help them appeal to existing or potential partners. These range from tattoos to jewels to dresses to houses to songs. The obsession of people with clothing themselves in ways that might be considered attractive is universal. It occupies a surprisingly large part of their consciousness and takes up a significant part of their budget: far too much to be explained by prosaic natural selection. Had I been writing this paragraph a century ago I would probably have said that females show more interest in dressing to impress than males. But today, as in much of history, I think this is no longer so true. Males may not be quite such 'peacocks' as females when it comes to fashion, preferring to signal wealth or originality in their clothing rather than elegance, but they still put a lot of thought into dressing so as to appeal to the opposite sex, especially around the age of pair formation.

Given that evidence from lice genes among other things points to hairless bodies having evolved many hundreds of thousands of years ago, it seems likely that people have been wearing clothes for a very long time, and probably decorating those clothes for most of it. At Sunghir, north-east of Moscow, a Paleolithic man who died around 32,000 years ago was buried with about 3,000 mammoth-ivory beads. These have ended up on the bones of his head, body and limbs, implying they were once part of extensive clothing. In Jordan, a Neolithic child was buried more than 8,000 years ago with a complex necklace made of 2,500 beads made from mother-of-pearl, amber and stone. In Iraq, a Bronze Age queen of Ur was buried around 4,500 years ago with more than seven pounds of gold and precious stones, gold rings, a gold-looped belt and an elaborate head-dress.

Sure, a lot of this decoration was signalling relative wealth. But why do people want to signal wealth? Partly at least to attract mates. The pursuit of profit, which has driven much of the progress in the world, was at root more likely to have been about sexual competition than survival. Once they have escaped poverty and immediate need, human beings do not slacken off their interest in seeking wealth or doing ambitious, creative things – they tend to accelerate them. Conspicuous consumption is a common phenomenon among the wealthy and one that is hard to explain without mentioning mating habits.

In particular, the pursuit of beauty, through art, is likely to have been selected over time not because it helped your ancestors get through the day but because it was more likely to catch the attention of a member of the opposite sex. At the very least that should be considered as a hypothesis. Geoffrey Miller, an evolutionary psychologist based in Albuquerque, New Mexico, is convinced that you can explain art a lot more easily through sexual selection theory than natural selection. As Ernst Grosse wrote in an 1897 book, *The Beginnings of Art*, 'Natural selection should long ago have rejected the peoples which wasted their force in so purposeless a way, in favour of other people of practical talents.' Grosse allowed that dance could have evolved through sexual selection but mostly he thought art evolved through 'strengthening and extension of the social bonds', which is no explanation at all: it is the thing that needs explaining. The same applies to the argument that art evolved because people enjoyed it, or that it was invented to appease the deities. These are not explanations.

Sexual selection cuts this Gordian knot. Miller points out that art is a signalling system, in which one individual signals something to others by manufacturing a signal, whether it be a painting, a poem or a pop song. Animals signal to each other all the time and when they do so it is much harder to figure out what is in it for the signaller than the signalled. That animals evolve ways to receive useful information from others' signals is unsurprising; that animals invest energy in sending signals is not so easy to understand. But when it comes to a bowerbird's bower, a Nightingale's song or a Black Grouse's display, it is obvious what is in it for the signaller: potential sex, potential posterity, potential genetic immortality. So Darwin argued, and – despite a century of doubt from others – I have explained why he was undoubtedly right.

Miller thinks that 'aesthetic admiration of art shades over into sexual attraction towards the artist – and this has been happening for several thousand generations. We're all descended from artists because art was sexy and art was romantic.' There are actual studies showing this to be true, statistically: one by Helen Clegg, Daniel Nettle and Dorothy Miell demonstrated that the degree of artistic success strongly correlated with mating success among male visual artists. This does not mean the sexual motivation of artists need be conscious or deliberate, or even that it is necessarily the common pattern today for artists of either sex to be sexually incontinent. All it would take to breed artistic creativity into the human species would be a small tendency towards sexual success for creative people at some point in the past. As with a Peacock's tail, it is possible that the force that created this instinct to be creative has faded away but the instinct remains. Sexual selection does not need to make its subjects conscious of how their habits emerged.

'Species capriciously transform themselves into their own sexual amusements'

Humour is another example. Making people laugh has minimal survival value. It does not often help you find food or avoid predators. But it demonstrably makes you popular, especially with the opposite sex. Both sexes highly value a good sense of humour in their partners, poll after poll confirms. In one survey by the American anthropologist Helen Fisher of a thousand people, 78 per cent of men and 84 per cent of women said it was 'very important' or 'somewhat important' that dates laugh at their jokes. Men seem to try harder to make their partners laugh and women seem to value being made to laugh more but there is plenty of overlap. Fisher found that when asked 'Have you ever fallen in love with someone because of their sense of humor?', 57 per cent of women said yes, and 40 per cent of men. Those are high numbers.

So suddenly in a world where mate choice drives evolution, the emergence of laughter, and a sense of humour, makes a lot more sense. It is one of many features of human behaviour that natural selection struggles to explain. As Miller put it in his book *The Mating Mind*, it is difficult to come up with natural selection arguments for things like storytelling, gossip, self-consciousness and ideology. But sexual selection through mate choice specialises in generating such 'apparently useless embellishments'. Natural selection is dull and utilitarian in its outputs; sexual selection is fun and strange. 'Species capriciously transform themselves into their own sexual amusements,' says Miller. Mate choice is therefore an excellent candidate to elucidate those 'luxury' features of human behaviour like religion, wit and art.

Moreover, sexual selection through mate choice is a two-way conversation between signaller and receiver. In Miller's colourful metaphor, 'like burglars learning about the security systems of banks, animals evolve courtship strategies to sneak through the senses of other animals, through the ante-chamber of their decision-making systems, into the vault of their reproductive potential'. And just as the banks change the locks, so animals evolve resistance to being seduced. When a Polar Bear adapts to life on the pack ice, the pack ice does not change in response. But while Peacocks evolve to impress females, Peahens evolve to be less easily impressed.

Yet to this day the vast majority of ideas about the evolution of the human mind invoke natural, not sexual selection. Where sexual selection is mentioned it is to explain differences between males and females, not between human beings and Chimpanzees. Popular theories to explain the rapid expansion of the human brain starting some three million years ago include the 'social brain' hypothesis, whereby human ancestors evolved big brains to keep track of relationships, plots and schemes within large and complex social groups. But explaining the unique expansion of the human brain using such natural selection explanations runs into the brick wall of a simple question: why did it not happen to other species? If it is so valuable to have extra grey matter, why don't all sorts of birds, monkeys and apes have it? Why are Bottlenose Dolphins the only creatures that come close to rivalling our brain size relative to body size? Sexual selection specialises in creating unique, runaway expansions found in no other species, so again it is a candidate that deserves consideration.

True Darwinists mostly shied away from considering sexual selection as a cause of the expansion of the human mind perhaps because it felt too much like intelligent design, and it gave up on the essential idea of evolution as an emergent phenomenon. Instead it suggests that deliberate choice by individuals is driving evolution. To admit this is to come far too close to falling back into creationism. Yet Darwin seems almost to have lost interest in natural selection once he had finished the *Origin of Species*, writing little about it and weakening his claims for it in successive editions of that book. By contrast, he wrote a lot about evolution by deliberate action – from the domestication of species to the role of insects in fertilisation and worms in decomposition to the role of females in sexual selection. He sounds almost like a proponent of intelligent design in the *Descent of Man* when he celebrates 'the remarkable conclusion that the cerebral system not only regulates the existing functions of the body, but has indirectly influenced the development of various bodily structures and of certain mental qualities'. This was in a sense far more heretical an idea than relying just on the passive forces of selective survival. Hence the fact that it encountered more than a century of determined rejection.

The neglect of sexual selection in the study of human beings has left anthropology, psychology, economics, political science and even social policy the poorer. Miller thinks there is a gap where the theory of sexual selection should be. From Marx to Freud to Veblen to Foucault to Chomsky, theories of human nature in the twentieth century were mostly failures, hollowed out by their neglect of possibly the most powerful of all evolutionary forces shaping the human mind, mate choice. As Miller argues,

> Economists could not explain our thirst for luxury goods and conspicuous consumption. Sociologists could not explain why men seek wealth and power more avidly than women. Education psychologists could not explain why students became so rebellious and fashion-conscious after puberty. Cognitive scientists could not fathom why creativity evolved. In each case apparent lack of 'survival value' made human behaviour appear irrational and maladaptive.

If this is so, why has the sexual selection theory of the human mind not so far spawned a rich school of enquiry? Miller's book was published in 2000 to good reviews, but most speculation about the enlargement of the human brain still focuses on natural not sexual selection.

A potential objection to a sexual selection explanation of the growth of the human brain is that it happened equally in both sexes. Crested Auklet males and females look almost identical, so sexual selection can be indifferent to sex, as I have said above. There are many species of brightly coloured birds in which both sexes are equally smart: parrots, penguins, kingfishers, bee-eaters, rollers and more. In Puffins both sexes have colourful beaks in the breeding season. (Usually these species nest in holes, as Wallace noted, removing the alternative need for camouflage in the female brooding the eggs.) But the pattern of preference in human beings would appear to include females selecting different features in males than those that males select in females.

This objection need not be fatal, however, because the brain could easily be a feature that both sexes valued almost equally in the past, when it was getting rapidly bigger. And quite often selection simply sees no need to impose a sex difference if there is no value to it. As Darwin argued: 'It is, indeed, fortunate that the law of the equal transmission of characters to both sexes has commonly prevailed throughout the whole class of mammals; otherwise it is probable that man would have become as superior in mental endowment to women, as the Peacock is in ornamental plumage to the Peahen.' To be clear, I am not endorsing Darwin's casual Victorian sexism here – his argument works just as well if you argue that men were after clever women more than vice versa, and yet ended up with cleverer children of both sexes. In any case, as Miller argues, the thing about brains is that they need to be clever both at transmitting and receiving. Making good jokes, and having a good enough sense of humour to appreciate them, are equally demanding of mental horsepower. If males are using their minds to try to seduce females, females could be using their minds to judge which male to choose, or vice versa.

Sexual attractiveness alone can be a sufficient explanation for almost any human mental trait. It should therefore be the null hypothesis. It's such a simple and obvious way for a feature to evolve that the burden of proof should be on alternative explanations to prove themselves. For example, people say they value kindness in potential mates. Look no further then for why human beings have a tendency to be kind. Sure, it might also be a spillover from the kindness towards your own children that results in you becoming a successful parent, or a form of enlightened self-interest that sees kindness as a down payment on the generosity of others. These are two good theories. But why is the mate-choice theory not just as plausible? Miller thinks so: 'Using the sturdy mule of mate choice to haul the cart of human nature up the mountain of morality ... Darwinians have searched so hard for the selfish survival benefits of morality that we have forgotten its romantic appeal.'

Knowing what we do about people's sexual preferences for kindness, moral behaviour would be bound to spread through mating preference.

The theory that the human mind evolved largely or partly through sexual selection by mate choice is radical. It may seem superficially to undermine the magic of romantic love; to tarnish the idea that great art is ethereal and eternal; to suggest that morality, humour and religion can be reduced to seduction devices; to imply that people do have heritable qualities, which is in itself a taboo idea, and are prepared to advertise these heritable qualities and seek them out in others, which sounds like dangerous talk. This is all a bit demeaning, but then science has specialised for centuries in taking human beings' view of themselves down a peg or two. Nicolaus Copernicus said the earth was not the centre of the universe. Charles Darwin said the human being was just another species of ape. Francis Crick found, to general surprise, that human beings used the exact same genetic code as a cabbage or a bacterium. The Human Genome Project revealed, contradicting conventional wisdom, that we have not only no special genes for building our brains, but that we have no more genes in total than a mouse. Get used to it. I believe that the human mind is probably one of sexual selection's greatest creations, and that mate choice is one of the most far-reaching ideas that the human mind can grapple with.

Epilogue
Saving the Black Grouse

Early January, the Pennine hills

Dawn comes late at this time of year, around 8.30 a.m. It's blowing a gale from the north-west laced with spits of high-speed sleet, the temperature not far above freezing. But as the light looms, the black shapes are there. Fifteen males are on the lek, each facing into the wind, hunkered down among the rushes. Occasionally one spreads his tail and utters a lager-can sneeze. A stand-off between two neighbours starts but does not last. The red combs above their eyes are shrunken, barely visible. It is three months before the next breeding season but every male feels the need to be there, to reserve his spot for the forth-coming campaign, to place his towel on the sun lounger for the show that will climax in April. Once more the dance moves will be compared, beauty will be judged and genes will be sieved.

This book began by asking why some male birds look so beautiful and act so extravagantly, with the lekking habit of the Black Grouse as a particular example. Why grow lyre-shaped tail feathers, bright white bum feathers, deep blue neck feathers, white spots on the flank and a pair of fake sea-anemones on the top of your head? Why gather in a tight flock and set up exclusive toy territories that you occupy every dawn for up to eight months of the year, fighting your neighbours over the boundaries again and again but never redrawing them? Why inflate your neck and bubble away for hours on end, interspersing this with loud sneezing sounds and flutter-jumps? It is a delightfully eccentric and yet gorgeous spectacle that defies rational explanation.

We can rule out the old idea that God has a sense of the beautiful (and the absurd), if only because that poses the further question, why? That would be turtles all the way down. The true answer has emerged from a century and a half of argument and evidence, ever since Charles Darwin and Alfred Wallace met at Down House with two friends to disagree on the matter in September 1868. Darwin was right: males do these things because females have bred them, selectively, to do these things, and done so gradually, over many thousands of generations. His critics were wrong to rule this out for a hundred years. Sexual selection through mate choice is a powerful evolutionary force, capable of much more eccentric outcomes than natural selection. Reproduction of the sexiest rather than survival of the fittest. That is still an underappreciated feature of evolution.

But why? The reason females did this to males – unconsciously, of course – was because they got carried away on a runaway train in a random direction. Yes, at various points in the process it probably also helped them choose healthy genes for their young and may still do so. But the real prize for them was not healthy offspring but attractive offspring. Once they started being choosy they had to go on getting more choosy, however random the criteria they were using in the choice, lest they fail to beget sexy sons. My main conclusion in this book is that the runaway and good-genes theories are not opposites and not in conflict. Both are true. A century and a half of debate, trying to distinguish between them, starting with Wallace versus Darwin, continuing with Huxley versus Fisher and more recently with Grafen versus Prum, can be laid to rest and resolved. Females drive males into adopting crazy but aesthetically pleasing plumage and behaviour because that way – again, unconsciously, of course – they will have offspring that are both slightly more likely to be healthy and slightly more likely to be attractive. Males gather at leks because females have bred them to do so as surely as dog breeders have bred poodles.

Yet a scientific explanation was only part of what I was seeking. To immerse myself in a natural phenomenon that is beautiful, and to convey to the reader its magic was just as big a part of my aim in writing this book. The lek of the Black Grouse is one of nature's most spectacular yet predictable spectacles. Almost every friend I have taken to watch a lek comes away thrilled by the experience. The species is Europe's 'bird of paradise'; its antics are unique, its colours unrivalled. Yet it is now vanishingly rare in Britain, where I live. Black Grouse once thrived in almost all the counties of England, plus most of Wales and Scotland – though it went extinct in Ireland a long time ago. The last national survey in 2019 found just 4,242 males breeding in Britain: 2,716 in Scotland, 1,281 in northern England and 245 in Wales. That is a 35 per cent decline since 2005. A good breeding season in the Pennines in 2022 and a reasonable one in 2023 has probably helped boost those numbers, but the danger that this bird could go extinct in Britain is real. And the British government, its conservation agencies and most of the conservation charities seem largely unconcerned.

The species is not in danger of global extinction, thriving throughout Scandinavia, Russia and parts of Eastern Europe and as far east as Inner Mongolia and Manchuria in eastern China – usually wherever there are birch trees. But throughout Eurasia, in most places Black Grouse are declining, while in the Alps and the Netherlands as in the British Isles they are now very rare. In Denmark the species died out in 2001. The problem is fairly simple: Black Grouse fare poorly under the impact of human activity, finding modern agriculture, modern forestry and the infrastructure of civilisation uncongenial in the extreme.

The decline of the Black Grouse was a long, slow retreat. As early as 1789, Gilbert White, the naturalist vicar of Selborne in Hampshire, wrote:

But there was a nobler species of game in this forest, now extinct, which I have heard old people say abounded much before shooting flying became so common, and that was the heath-cock, black-game, or grouse. When I was a little boy I recollect one coming now and then to my father's table. The last pack remembered was killed about thirty-five years ago; and within these ten years one solitary greyhen was sprung by some beagles in beating for a hare.

The species seems to have done better in the 1800s, reappearing in good numbers in some areas where seed-rich cereal fields and dung-rich cattle pastures mixed with heath and scrubby woods provided the sort of varied habitat the Black Grouse loves. But after the First World War it was again in decline, soon disappearing from the lowlands altogether.

It was dealt hammer blows even in the uplands by two government policies in the second half of the twentieth century. First there was a policy of subsidising commercial forestry in the futile, mercantilist hope of reducing Britain's dependence on imported timber. A large, nationalised business called the Forestry Commission began buying huge tracts of land to plant with exotic conifers, mainly Sitka spruce, while also pushing tax breaks towards private landowners who did the same. At first these new forests delighted the Black Grouse, which thrived in the rich, ungrazed grass and weeds of the young plantations. The foresters even treated Black Grouse as pests, on the grounds that the birds ate the tips of growing trees, causing the tree to grow less straight. They put a bounty on their heads in the 1950s. But as soon as the trees grew up, and turned into dark, closed-canopy woods, the Black Grouse starved and vanished. Throughout south-west Scotland, north Northumberland and Wales, in particular, where Black Grouse used to be numerous, the species has now almost entirely gone extinct thanks to commercial forestry. The Forestry Commission was even more zealous than private landowners in the way it converted open, messy woods of birch and oak – perfect for Black Grouse – into regimented lines of dark, dense, even-aged spruce.

The other government policy that harmed the species was 'improvement' of the farms in the uplands. Black Grouse and sheep get on well, but not if the latter are too densely stocked. The birds avoid areas that are heavily grazed especially in winter since this removes the herbs they love to eat and the thicker vegetation in which they like to roost. Subsidies for sheep grazing encouraged larger flocks and more intense grazing, leading to the loss and degradation of vast tracts of heather moorland, while drainage schemes shrank the rushy bogs where they also liked to feed, roost and breed, and created ditches that trapped chicks. Subsidies for intensive farming encouraged the replacement of hay with silage, which is an unpalatable monoculture of rye grass, and is harvested early before any flowers have bloomed and set seed – and while birds are still sitting on nests. The hills of northern Britain and Wales had once sported a patchwork quilt of moors, bogs, open woods, scrubby bushes, oat fields and hay meadows, which gave the Black Grouse something for every season. But now it has become mostly a clean-cut, black-and-white mixture of too-dark woods and too-green fields, neither of which is any good for the birds.

At the same time the relentless advance of human activity brought with it a new threat to the birds: predators. Whereas before Black Grouse had lived with predation by hawks, crows, stoats and foxes, now human activity increasingly provides predictable food for predators in the form of road-kill, landfill, refuse, agricultural feed and other sources. But instead of distracting the predators, this unnatural and unseasonal food supply subsidises their populations, enabling them to grow their numbers and hold populations higher through the winter ready to devastate nests and chicks of Black Grouse in the spring. Gulls and rats, thriving on human refuse, have also spread into the uplands in the same way. This 'subsidising' of predators is a well-known

phenomenon in conservation, where human activity enables a predator to live at higher densities through lean seasons, resulting in a
devastating effect on its natural prey. For instance, a sharp decline in
desert tortoise numbers in a California desert was traced to the opening of a landfill site nearby, resulting in a sharp increase in Raven
numbers. Making the problem worse, Britain has lost almost all its
'apex' predators, like eagles, wolves and lynx, which would once have
controlled the number of generalist 'meso-predators' like foxes, badgers, Buzzards and crows. In an ecosystem it often proves to be the case
that killing off the top dog allows the middle dog to devastate the
bottom dog.

Yet throughout the time the Black Grouse has declined, the Red
Grouse – a bird unique to the British Isles – has thrived. Its numbers
fluctuate from year to year and decade to decade, and it has vanished
from most of Ireland, Devon and Cornwall, Wales and parts of
Scotland, but in the Pennines and eastern Highlands of Scotland, it
continues to live at extraordinarily high densities, unparalleled for a
wild bird in a low-productivity habitat. There are probably at least fifty
Red Grouse in Britain today for every Black Grouse. On moors dominated by heather, a plump bird that converts heather into meat with
great efficiency can now be found living at a pair to the acre or more,
and they are a much valued delicacy.

This was unexpected. The farmer and writer Patrick Laurie found a
remark in the 1860 edition of *Encyclopaedia Britannica* predicting that
the Black Grouse would adapt better than the Red Grouse to the
advance of civilisation. 'Red Grouse recede where civilisation
progresses,' the encyclopaedia wrote, 'and they are consequently, in a
fair way, at no very distant period, of being banished from England. As
the vast extent of heath-land is not required for black grouse, there is
no room to fear their extinction for some centuries to come.' In fact,
though its diet does vary a little, the Red Grouse's essential needs are
simple – heather, heather and heather – while the Black Grouse needs

tree buds in the winter, herbs in the spring, seeds and berries in the autumn and insects for chicks in the summer. 'In the last century, we have knocked his entire dietary calendar into chaos,' as Laurie puts it. .

The reason for the Red Grouse's remarkable success lies with human beings too. In the 1800s the sport of shooting Red Grouse became popular and hence valuable, so the owners of moors began to manage the land with grouse in mind as well as, or instead of, sheep. This meant burning small patches of heather to encourage fresh shoots, trapping foxes, stoats and crows to reduce predation and providing grit to help these birds grind up and ferment their diet of heather shoots. Rearing grouse never proved possible, so the population of this quarry species remains entirely wild, but the effect was a dramatic increase in the density of Red Grouse on heather moors. In the 1800s, gamekeepers on grouse moors also killed eagles, hawks, harriers, owls and falcons – and then the use of DDT in the 1950s and 1960s proved unintentionally even more lethal to these raptors. Today there is a war of words between some conservationists who emphasise that such killing occasionally continues and other conservationists who point out that the populations of these raptors are recovering rapidly to levels not seen for centuries. Hen Harriers for example have gone from zero chicks raised in England in 2013 to 141 in 2023. Their revival is threatening the livelihoods of gamekeepers, by devastating grouse populations where they gather to breed in semi-colonial fashion, creating a classic human-wildlife conflict, not unlike what happens when elephants raid farmers' fields.

What is not in doubt is that the moors managed for grouse came to be ideal habitat for many other birds, animals and plants. Curlews, Golden Plover, Merlins, lizards, Emperor Moths, a dozen varieties of sphagnum mosses, and many other species thrive to this day on these managed grouse moors, providing an impressive experiment in what is effectively privately funded conservation. This creates and protects a rare habitat largely confined to the British Isles that is uncultivated, unsprayed and unfertilised so it remains 'wild' in a lightly managed way. Open, treeless moorland is also much more natural than we once thought – the result of wet summers following a climatic cooling around six thousand years ago more than human clearing and grazing as was once thought. In the Pennines, the management of moorland for grouse kept at bay the commercial forestry, the wind farms and the intensive grazing that would have swept away such species. It was this management for Red Grouse that has also saved the Black Grouse so far. Almost all – around 90 per cent – of the Black Grouse that remain in England live on the edges of grouse moors, where landowners control the sheep numbers and gamekeepers control the fox, stoat and crow numbers. This is their last refuge.

The reliance on shooting or hunting for saving a species is an uncomfortable reality for many conservationists, but the truth is that saving the Black Grouse without hunting has been and is being tried in both Wales and Scotland by the Royal Society for the Protection of Birds and it is failing. The habitat is being managed well for birds but without relentless predator control, the Black Grouse population dwindles. At Lake Vyrnwy in Wales, Black Grouse numbers on RSPB-owned land are now dangerously low and in the organisation's own words, 'in the next few years', the birds will 'cease to appear as a breeding species' unless something is done. In 2023 there were just three males left there. The species is also declining on the RSPB's land in Speyside in Scotland. In the Pennines, where grouse moors thrive, numbers are stable and the range of the bird is at last expanding again. It is legal to shoot Black Grouse still, but nobody does so deliberately, and on most shooting estates if one is shot by accident a hefty fine is levied and paid to a Black Grouse conservation project.

Finland teaches the same lesson. Black Grouse thrive there but only because the hunters who like to shoot and eat them (in their hundreds of thousands) also control their predators including pine martens and badgers. Hunters are the birds' best friends. The bald truth is that this species cannot thrive alongside human beings unless human beings intervene to reduce the populations of subsidised predators as well as to manage the habitat, and that cannot be done at scale except by hunters. As in southern Africa, where hunting preserves larger tracts of land for wildlife than national parks do, the best hope is if conservationists and hunters work together.

'Rewilding' in its purest sense will therefore not help the Black Grouse. The species does not like dense woodland or rank vegetation and it cannot cope with modern levels of subsidised predation that come with a hands-off approach. What it needs is 'managed wilding': the deliberate creation or maintenance of a mosaic pattern of bog, moor, meadow and open birch woodland, together with careful predator control. Grouse in general are birds with a specialised talent for surviving northern winters by eating twigs or bugs of woody plants, fermenting them in the caecum, a special part of the gut. Each species has adapted to a staple woody plant to help it get through the winter, as often shown by the very name of the species: Sage, Hazel, Willow, Spruce Grouse. Red Grouse should really be called Heather Grouse and Capercaillies Pine Grouse. The Black Grouse is known as Birkhuhn ('birch hen') in Germany: it is reliant on the buds and catkins of birch trees in most of its range to get it through the winter, yet in Britain there is no government policy to encourage foresters to plant birch instead of alien spruce. Indeed, sadly I hear few hints of interest in Black Grouse from Natural England, the government's all-powerful conservation agency. Managed wilding would benefit many other species too, including human beings who could farm and manage such land.

If British environmental policy fails to pursue these policies, it will be responsible for possibly, probably, committing our 'bird of paradise' to national extinction. I desperately hope this does not happen, and it is one of the reasons I focused this book to celebrate the extraordinary lek of the Black Grouse. If and when the leks in Britain fall silent, the lager-can sneezes and bubbling roo-koos fade on the dawn wind and the black-and-white dancing ceases, it will be a tragedy.

Acknowledgements

As I researched and wrote this book, I experienced great generosity from all those to whom I reached out for help, both practical and intellectual, literary and scientific. In the Pennines, my many visits to Black Grouse leks are made possible by friends and allies, among them especially Mark Gallagher and Will Makepiece, and my semi-regular lek-watching companions: Jamie Blackett, Willy Browne-Swinburne, Matthew Dryden, James Fenwick, Emily Graham, Reggie Heyworth, David Hill, Andrew Hopetoun, Tarquin Millington-Drake, William Morrison-Bell, Ralph Northumberland, Tim Palmer, Mark Tellwright and Christopher Wills. I have also shared dawn lek sessions and discussions with other friends from my days as a biologist, including Mike Birkhead, Tim Birkhead, Jonathan Kingdon, Philip Merricks, Ian Newton, Chris Perrins and Mike Rands. I thank Matthew Dryden, David Hill, Jonathan Pointer, James Boettcher and Henry Koh for the use of their photographs. Dave Baines and Phil Warren generously shared their insight into the ecology of Black Grouse in Britain, as did Ian Coghill. For responding to my queries about various aspects of sexual selection theory and experiments, I thank Andrew Balmford, Mark Boyce, Raymond Corbey, Helena Cronin, Geoffrey Davison, Mercedes Foster, Alan Grafen, Nigella Hillgarth, Ian Jones, Clemens Küpper, Russ Lande, Dov Lank, Justin Marshall, Geoffrey Miller, Dustin Penn, Marion Petrie, Richard Prum, Evelleen Richards and Chris Thouless. For sharing his remarkable research on Finnish Black

Grouse, I thank Carl Soulsbury. In Norway, I am grateful for the help of Tormod Amundsen and Terje Kolaas; in Australia, James Boettcher; in Papua New Guinea, Logan Tube. My agent Georgina Capel and my editors Louise Haines and Terry Karten championed this book when others were doubtful. Thanks too for the fine copy-editing skills of Iain Hunt. My best adviser, as always, is my wonderful wife Anya, who knows more about colour than anybody alive.

Notes

1. The Lek

3 'The word is Swedish, meaning "play"'. Höglund, J. and Alatalo, R.V. 1995. *Leks.* Princeton University Press.

6 'The impossibly wealthy heiress Mary Eleanor Bowes'. Moore, W. 2010. *Wedlock: The true story of the disastrous marriage and remarkable divorce of Mary Eleanor Bowes, Countess of Strathmore.* Crown.

11 'Some of the rivals on this lek will probably be brothers'. Lebigre, C., Alatalo, R.V., Soulsbury, C.D., Höglund, J. and Siitari, H. 2014. Limited indirect fitness benefits of male group membership in a lekking species. *Molecular Ecology* 23:5356–5365.

11 'This puzzle has been known to biologists since 1979'. Borgia, G. 1979. In Blum, M.S. and Blum, N.A. *Sexual Selection and Reproductive Competition in Insects.* Academic Press.

16 'As William Yarrell put it in 1843'. Yarrell, W. 1843. *A History of British Birds.* John van Voorst.

16 'The Newcastle engraver Thomas Bewick'. Bewick, T. 1797–1804. *A History of British Birds.* Longman.

18 'The artist John Guille Millais'. Millais, J.G. 1894. *Game Birds and Shooting Sketches.* H. Sotheran.

18 'The Northumbrian naturalist Abel Chapman'. Chapman, A. 1924. *The Borders and Beyond: Arctic, Cheviot, Tropic.* Gurney and Jackson.

19 'Two Dutch scientists observed the same thing'. Kruijt, J.P. and Hogan, J.A. 1967. Social behavior on the lek in Black Grouse, *Lyrurus tetrix tetrix* (L.). *Ardea* 55:204–240.

20 'Llewellyn Lloyd, a British sportsman'. Lloyd, L. 1867. *The Game Birds and Wild Fowl of Sweden and Norway: with an account of the seals and salt-water fishes of those countries.* F. Warne.

20 'It came from Edmund Selous'. Selous, E. 1909–10. An observational diary on the nuptial habits of the blackcock (*Tetrao tetrix*) in Scandinavia and England. *Zoologist* 13: 401–413 and 14: 23–29, 51–56, 176–182, 248–265. Reprinted in Selous, E. 1927. *Realities of Bird Life: being extracts from the diaries of a life-loving naturalist.* London: Constable.

24 'It was Rice University's David Queller'. Queller, D.C. 1987. The evolution of leks through female choice. *Animal Behaviour* 35:1424–1434.

25 'David Lack, a professional academic zoologist'. Lack, D. 1939. The display of the blackcock. *British Birds* 32:290–303.

25 'A young Swiss-born medical scientist named Otto Hohn'. Hohn, E.O. 1953. Display and mating behaviour of the Black Grouse *Lyrurus tetrix* (L.). *British Journal of Animal Behaviour* 1:48–58.

26 'At Loch Ard in Scotland in the 1950s Kenneth Richmond'. Richmond, W.K. 1963. A note on the communal display of Blackcocks. In Bannerman, D.A. *The Birds of the British Isles.* Vol. 12. Oliver and Boyd.

26 'Between 1960 and 1963, Ilkka Koivisto'. Koivisto, I. 1965. Behaviour of the black grouse, *Lyrurus tetrix* (L.), during the spring display. *Finnish Game Research.*

28 'In the Netherlands between 1961 and 1964 J.P. Kruijt and Jerry Hogan'. Kruijt, J.P. and Hogan, J.A. 1967. Social behavior on the lek in Black Grouse, *Lyrurus tetrix tetrix* (L.). *Ardea* 55:204–240.

30 'A study in Finland found that eye combs'. Harris, S., Kervinen, M., Lebigre, C., Pike, T.W. and Soulsbury, C.D. 2018. Age, condition and dominance-related sexual ornament size before and during the breeding season in the black grouse *Lyrurus tetrix. Journal of Avian Biology* 49: e01648.

31 'A follow-up study by the Finland team'. Harris, S., Kervinen, M., Lebigre, C., Pike, T.W., Soulsbury, C.D. and Herberstein, M. 2020. Full spectra coloration and condition-dependent signaling in a skin-based carotenoid sexual ornament. *Behavioral Ecology* 3:834–843.

2. Darwin's Unpopular Theory

33 'He wrote in September 1828'. Darwin's correspondence and notebooks can be read online at darwinproject.ac.uk.

34 'In his book *The Descent of Man* in 1871'. Darwin, C.R. 1871. *The Descent of Man and Selection in Relation to Sex.* John Murray.

36 'The historian of science Evelleen Richards'. Richards, E. 2017. *Darwin and the Making of Sexual Selection.* Chicago University Press.

36 'In 1794 in his book *Zoonomia, or the Laws of Organic Life*'. Darwin, E.R. 1794–6. *Zoonomia; or the Laws of Organic Life.* J. Johnson.

39 'In 1844 Darwin set out his idea'. Hoquet, T. and Levandowsky, M. 2015. Utility vs beauty: Darwin, Wallace and the subsequent history of the debate on sexual selection. In Hoquet, T. (eds) *Current Perspectives on Sexual Selection. History, Philosophy and Theory of the Life Sciences*, Vol. 9. Springer, Dordrecht.

40 'In the first edition of the *Origin of Species* in 1859'. Darwin, C.R. 1859. *On the Origin of Species.* John Murray.

40 'His friend and mentor in the pigeon-fancying world, John Eaton'. Eaton, J. 1858. *A treatise on the art of breeding and managing tame, domesticated, foreign, and fancy pigeons, carefully compiled from the best authors, with observations and reflections, containing all that is necessary to be known of tame, domesticated, foreign and fancy pigeons, in health, disease, and their cures.* London.

42 'The eighth Duke of Argyll'. Argyll, the Duke of. 1867. *The Reign of Law.* A. Strahan.

46 'Wallace wrote a quarter of a century later'. Wallace, A.R. 1889. *Darwinism.* Macmillan.

47 'As Helena Cronin spotted'. Cronin, H. 1991. *The Ant and the Peacock.* Cambridge University Press.

49 'A short letter in *Nature* from a butterfly collector, George Fraser'. Fraser, G. 1871. Sexual selection. *Nature* 3:489.

50 'A young man named Thomas Wood'. Wood, T.W. 1870–71. The courtship of birds. *The Student and Intellectual Observer of Science, Literature and Art* 5:113–125.

53 'An anonymous reviewer of the *Descent of Man*'. Anon. 1871. Review of *The Descent of Man. British Quarterly Review* 53:565–569.

54 'The explorer William Beebe'. Beebe, W. 1918–22. A *Monograph of the Pheasants.* Witherby.

55 'The dedicated bird photographer Tim Laman'. Laman, T. 2022. How we filmed the Great Argus Pheasant. Timlaman.com, Wildlife Diaries, February 8, 2022.

55 'As Richard Prum puts it'. Prum, Richard O. 2017. *The Evolution of Beauty.* Knopf Doubleday Publishing Group.

56 'In 2022 two scientists at the University of Exeter's Cornwall campus'. Firkins, L.M.E. and Kelley, J.A. 2022. Does shading on Great Argus *Argusianus argus* feathers create a three-dimensional illusion? *Biological Letters* 18:20220393.

61 'When Alfred Russel Wallace came to review the *Descent of Man*'. Wallace, A.R. 1871. [Review of] Darwin's 'The Descent of Man and Selection in Relation to Sex'. *The Academy* 2:177–182. 15 March 1871.

62 'St George Mivart was a lawyer turned philosopher'. Mivart, St. G.J. 1871. [Review of] The Descent of Man, and selection in relation to sex. *Quarterly Review* 131:47–90.

63 'Nor was Herbert Spencer'. Spencer, H. 1890. The origin of music. *Mind* 15:449–468.

3 The Females Arrive

70 'Black spots at the tips of feathers'. Soulsbury, C.D., Kervinen, M. and Lebigre, C. 2016. Curse of the black spot: spotting negatively correlates with fitness in Black Grouse *Lyrurus tetrix*. *Behavioral Ecology* 27:1362–1369.

72 'The meticulous Dutch study by J.P. Kruijt and Jerry Hogan'. Kruijt, J.P. and Hogan, J.A. 2015. Social behavior on the lek in Black Grouse, *Lyrurus tetrix tetrix* (L.). *Ardea* 55:204–240.

75 'First spotted by Edmund Selous'. Selous, E. 1909–10. An observational diary on the nuptial habits of the blackcock (*Tetrao tetrix*) in Scandinavia and England. *Zoologist* 13: 401–413 and 14: 23–29, 51–56, 176–182, 248–265. Reprinted in Selous, E. 1927. *Realities of Bird Life: being extracts from the diaries of a life-loving naturalist.* London: Constable.

76 'He wrote a furious note in the Yorkshire publication'. Selous, E. 1913. The nuptial habits of the blackcock. *Naturalist* 1913:96–98.

76 'In the latest edition of *The British Bird Book*'. Kirkman, F.B. and Jourdain, F.C.R. 1910–1912. *The British Bird Book: an account of all the birds, nests and eggs found in the British Isles.*

76 'By 1877 in an article in *Macmillan's* magazine'. Wallace, A.R. 1877. The colours of animals and plants: I and II. *Macmillan's* 36:384–408 and 464–471.

80 'Conwy Lloyd Morgan, a student of Thomas Henry Huxley'. Morgan, C.L. 1890. *Animal Life and Intelligence.* E. Arnold.

81 'Romanes argued that Wallace was missing the point'. Romanes, G. 1893. *Darwin, and After Darwin.* Longmans, Green and Co.

82 'The travel writer W.H. Hudson waxed lyrical'. Hudson, W.H. 1892. *The Naturalist in La Plata.* Chapman and Hall.

82 'A close friend of Joseph Conrad named Norman Douglass'. Douglass, G.N. 1895. On the Darwinian hypothesis of sexual selection. *Natural Science* 7:326–332, 398–406.

82 'An influential biologist named Henry Eliot Howard'. Howard, H.E. 1903. On sexual selection and the aesthetic sense in birds. *Zoologist* 7:407–417.

85 'Haven Wiley watched three Sage Grouse leks'. Wiley, R.H. 1973. Territoriality and non-random mating in Sage Grouse, *Centrocercus urophasianus. Animal Behaviour Monographs* 6:85–169.

88 'Rauno Alatalo and his colleagues repeated the experiment'. Höglund, J., Alatalo, R.V., Gibson, R.M. and Lundberg, A. 1995. Mate-choice copying in Black Grouse. *Animal Behaviour* 49:1627–1633.

89 'In an experimental facility near Laramie, Wyoming'. Spurrier, M.F., Boyce, M.S. and Manly, B.F.J. 1994. Lek behaviour in captive Sage Grouse *Centrocercus urophasianus. Animal Behaviour* 47:303–310.

4. Runaway Fashion

93 'Thomas Hunt Morgan'. Morgan, T.H. 1903. *Evolution and Adaptation.* Macmillan.

94 'The Scottish biologist and mathematician Sir D'Arcy Thompson'. Thompson, D.W. 1917. *On Growth and Form.* Cambridge.

94 'August Weismann gave a positive assessment'. Weismann, A. 1904. *The Evolution Theory.* British edition, Edward Arnold.

95 'As the philosopher David Rothenberg put it'. Rothenberg, D. 2012. *Survival of the Beautiful.* Bloomsbury.

95 'The German philosopher and psychologist Karl Groos'. Groos, K. 1898. *The Play of Animals.* D. Appleton and Co.

96 'The Polish ornithologist Jean Stolzmann'. Quoted in Hiraiwa-Hasegawa, M. 2000. The sight of the peacock's tail makes me sick: The early arguments on sexual selection. *Journal of Biosciences* 25:11–18.

96 'Patrick Geddes and J. Arthur Thomson'. Geddes, P. and Thomson, J.A. 1889. *The Evolution of Sex.* W. Scott.

99 '*The Making of Species* was published in 1909'. Dewar, D. and Finn, F. 1909. *The Making of Species.* John Lane, Bodley Head.

99 'In a letter written in the 1930s, Fisher wrote'. Fisher's letters are at digital. library.adelaide.edu.au.

100 'A short paper entitled "The Evolution of Sexual Preference"'. Fisher, R.A. 1915. The evolution of sexual preference. *Eugenics Review* 7:184–192.

101 'In his book *The Genetical Theory of Natural Selection*'. Fisher, R. 1930. *The Genetical Theory of Natural Selection*. Oxford.

104 'As Richard Prum argues'. Prum, Richard O. 2017. *The Evolution of Beauty*. Knopf Doubleday Publishing Group.

107 'Reported Miller in his book *The Mating Mind*'. Miller, G. 2000. *The Mating Mind*. William Heinemann.

108 'Ronald Fisher exchanged letters with a physicist friend'. Henshaw, J.M. and Jones, A.G. 2019. Fisher's lost model of runaway sexual selection. *Evolution* 74:487–494.

110 'He went so far as to claim in his paper on divers'. Huxley, J.S. 1923. Courtship activities in the Red-throated Diver (*Colymbus stellatus* Pontopp.); together with a discussion of the evolution of courtship in birds. *Zoological Journal of the Linnean Society* 35:253–292.

111 'An 895-page treatise on *The Science of Life*'. Wells, H.G., Huxley, J.S. and Wells, G.P. 1929. *The Science of Life*. Cassells.

111 'A paper by Huxley and C.P. Blacker'. Blacker, C.P. 1931. The sterilization proposals. *Eugenics Review* 22:239–247.

111 'As Huxley said in his Galton Lecture in 1936'. Huxley, J.S. 1936. Eugenics and society: the Galton lecture. Published in the *Eugenics Review* 28:11–31.

112 'In 1938 Huxley wrote a long essay for the *American Naturalist*'. Huxley, J.S. 1938. Darwin's theory of sexual selection and the data subsumed by it, in the light of recent research. *American Naturalist* 72:416–433.

113 'Hingston published a book'. Hingston, R.W.G. 1933. *The Meaning of Animal Colour and Adornment*. E. Arnold.

113 'Oskar Heinroth's description of the spectacular display'. Heinroth, O. 1938. Die Balz des Bulwersfasans, *Lobiophasis bulweri* Sharpe. *Journal für Ornithologie*. January 1–4.

5. Long Odds

119 'Occasional "Rackelhahns"'. See 'Grousing about the Rackelhahn – a spectacularly cross-tempered crossbreed of the feathered variety!' Karlshuker.blogspot.com, 23 Nov 2014.

120 'In Finland, after filming 1,082 fights'. Hämäläinen, A., Alatalo, R.V., Lebigre, C. et al. 2012. Fighting behaviour as a correlate of male mating success in black grouse *Tetrao tetrix*. *Behavioral Ecology and Sociobiology* 66:1577–1586.

122 'Back in 1991 Alatalo and his colleagues'. Alatalo, R., Höglund, J. and Lundberg, A. 1991. Lekking in the black grouse: a test of male viability. *Nature* 352:155–156.

124 'As Kervinen, Lebigre and Soulsbury write'. Kervinen, M., Lebigre, C. and Soulsbury, C.D. 2016. Simultaneous age-dependent and age-independent sexual selection in the lekking black grouse (*Lyrurus tetrix*). *Journal of Animal Ecology* 85:715–725.

125 'In a paper the Finnish team published in 2015'. Kervinen, M., Lebigre, C., Alatalo, R., Siitari, H. and Soulsbury, C. 2015. Life-history differences in age-dependent expressions of multiple ornaments and behaviors in a lekking bird. *American Naturalist* 185:13–27.

126 'In Finland 13 out of 193 yearlings did manage to mate'. Kervinen, M., Alatalo, R.V., Lebigre, C., Siitari, H. and Soulsbury, C.D. 2012. Determinants of yearling male lekking effort and mating success in black grouse (*Tetrao tetrix*). *Behavioral Ecology* 23:1209–1217.

127 'Another paper by Carl Soulsbury and his colleagues'. Nieminen, E., Kervinen, M., Lebigre, C. and Soulsbury, C.D. 2016. Flexible timing of reproductive effort as an alternative mating tactic in black grouse (*Lyrurus tetrix*) males. *Behaviour* 153:927–946.

129 'In 1988 Bruce Beehler and Mercedes Foster'. Beehler, B.M. and Foster, M.S. 1988. Hotshots, hotspots, and female preference in the organization of lek mating systems. *American Naturalist* 131:203–219.

130 'In a study of a Sage Grouse lek in Wyoming'. Snow, S.S., Patricelli, G.L., Butts, C.T., Krakauer, A.H., Perry, A.C., Logsdon, R.M. and Prum, R.O. 2022. Fighting isn't sexy in lekking Greater Sage-grouse (*Centrocercus urophasianus*). *bioRxiv*.

132 'In the coastal forests of Ecuador lives a bird'. Anderson, H.L., Olivo, J. and Karubian, J. 2023. The adaptive significance of off-lek sociality in birds: A synthetic review, with evidence for the reproductive benefits hypothesis in Long-wattled Umbrellabirds. *Ornithology* 140:1–21.

133 'Richard Prum has described how pairs of male Golden-winged Manakins'. Prum, Richard O. 2017. *The Evolution of Beauty*. Knopf Doubleday Publishing Group.

135 'Most male Black Grouse had no closely related neighbour on a lek'. Lebigre, C., Timmermans, C. and Soulsbury, C. 2016. No behavioural response to kin competition in a lekking species. *Behavioral Ecology and Sociobiology* 70:1457–1465.

136 'The Giant Elk's antlers are just about the "right" size'. Klinkhamer, A.J., Woodley, N., Neenan, J.M., Parr, W.C.H., Clausen, P., Sánchez-Villagra,

M.R., Sansalone, G., Lister, A.M. and Wroe, S. 2019. Head to head: the case for fighting behaviour in *Megaloceros giganteus* using finite-element analysis. *Proceedings Biological Sciences* 286:1912.

136 'Professor Adrian Lister of London's Natural History Museum'. Askham, B. and Hendry, L. The Irish elk: when and why did this giant deer go extinct and what did it look like? Natural History Museum website NHM.ac.uk.

139 'When displaying it jumped once every sixty-one seconds'. Ridley, M.W., Magrath, R.D. and Woinarski, J.C.Z. 1985. Display leap of the lesser florican *Sypheotides indica*. *Journal of the Bombay Natural History Society* 82:271–277.

139 'This small size would not help males in fights'. Raihani, G., Székely, T., Serrano-Meneses, M.A., Pitra, C. and Goriup, P. 2006. The influence of sexual selection and male agility on sexual size dimorphism in bustards (Otididae). *Animal Behaviour* 71:833–838.

6. Tales of Long Tails

141 'O'Donald was studying a seabird in the Shetland Islands'. O'Donald, P. 1972. Sexual selection for colour phases in the Arctic Skua. *Nature* 238:403–404.

141 'In 1962 O'Donald published the first mathematical models'. O'Donald, P. 1962. The theory of sexual selection. *Heredity* 17:541–552.

142 'His name was Robert Trivers'. Trivers, R.L. 1972. Parental investment and sexual selection. In *Sexual Selection and the Descent of Man, 1871–1971*. Editor, B. Campbell. Aldine, Chicago.

144 'The bird I was watching is a fabulously unusual species'. Ridley, M.W. 1980. The breeding behaviour and feeding ecology of Grey Phalaropes *Phalaropus fulicarius* in Svalbard. *Ibis* 122:210–226.

145 'As Alfred Russel Wallace wrote in 1870'. Wallace, A.R. 1870. *Contributions to the Theory of Natural Selection: a series of essays.* Macmillan.

150 'The University of Chicago's Russell Lande'. Lande, R. 1981. Models of speciation by sexual selection on polygenic traits. *Proceedings of the National Academy of Sciences of the United States of America* 78:3721–3725.

151 'Another mathematical model done by Mark Kirkpatrick'. Kirkpatrick, M. 1982. Sexual selection and the evolution of female choice. *Evolution* 36:1–12.

153 'On a visit to Kenya, Andersson had a better idea'. Andersson, M. 1982. Female choice selects for extreme tail length in a widowbird. *Nature* 299: 818–820.

156 'Andersson reflected on how the experiment had changed his life'. Sridhar, H. 2018. Reflections on papers past: revisiting Andersson 1982. *Rapid Ecology*, 17 October 2018.

156 'The stalk-eyed fly, *Cyrtodiopsis dalmanni*'. Wilkinson, G.S. and Reillo, P.R. 1994. Female choice response to artificial selection on an exaggerated male trait in a stalk-eyed fly. *Proceedings of the Royal Society of London B* 255:1–6.

157 'Andrew Pomiankowski and his colleagues'. Cotton, A.J., Földvári, M., Cotton, S. and Pomiankowski, A. 2014. Male eyespan size is associated with meiotic drive in wild stalk-eyed flies (*Teleopsis dalmanni*). *Heredity* 112:363–369.

158 'A species of small, blood-sucking sand fly'. Jones, T.M. 1997. Sexual selection in the sandfly *Lutzomyia longipalpis*. PhD thesis, London School of Hygiene and Tropical Medicine and Institute of Zoology.

159 'And Darwin-Fisher-sexy-son won, hands down'. Jones, T.M., Quinnell, R.J. and Balmford, A. 1998. Fisherian flies: benefits of female choice in a lekking sandfly. *Proceedings of the Royal Society of London B* 265:1651–1657.

7 Curlew Chorus

160 'Patrick Laurie put it in his book *Native*'. Laurie, P. 2020. *Native: life in a vanishing landscape.* Birlinn.

163 'As the great African scientist-artist Jonathan Kingdon put it'. Kingdon, J. 2023. *Origin Africa: safaris in deep time.* Princeton University Press.

164 '171 chicks from 52 broods of Skylarks were genotyped'. Hutchinson, J.M.C. and Griffith, S.C. 2007. Extra-pair paternity in the Skylark *Alauda arvensis. Ibis* 150:90–97.

164 'In two different populations of Curlews in Finland'. Currie, D. and Valkama, J. 2000. Population density and the intensity of paternity assurance behaviour in a monogamous wader: the Curlew *Numenius arquata. Ibis* 142:372–381.

167 'Dag Eriksson and Lars Wallin put out nest boxes'. Eriksson, D. and Wallin, L. 1986. Male bird song attracts females – a field experiment. *Behavioral Ecology and Sociobiology* 19:297–299.

167 'Clive Catchpole did a series of experiments with Sedge Warblers'.
Catchpole, C. and Slater, P.J.B. 2008. *Bird Song: biological themes and variations.* Cambridge.

169 'A study that finds that playing the sexy syllables'. Leitner, S., Voigt, C., Metzdorf, R. and Catchpole, C.K. 2005. Immediate early gene (ZENK, Arc) expression in the auditory forebrain of female canaries varies in response to male song quality. *Journal of Neurobiology* 64:275–284.

172 'Male Crested Auklets are almost identical to females'. Jones, I. and Hunter, F. 1993. Mutual sexual selection in a monogamous seabird. *Nature* 362:238–239.

175 'Nearly sixty miles an hour'. Klaassen, R.H.G., Alerstam, T., Carlsson, P., Fox, J.W. and Lindström, Å. 2011. Great flights by great snipes: Long and fast non-stop migration over benign habitats. *Biological Letters* 7:833–835.

178 'The male Great Snipe is perfectly camouflaged'. Höglund, J., Kalis, J.A. and Løfaldli, L. 1990. Sexual dimorphism in the lekking great snipe. *Ornis Scandinavica* 21:1–6.

180 'Höglund and his colleagues did an experiment'. Höglund, J., Eriksson, M. and Lindell, L.E. 1990. Females of the lek-breeding great snipe, *Gallinago media*, prefer males with white tails. *Animal Behaviour* 40:23–32.

180 'He removed certain male Great Snipe from the lek for one night'. Höglund, J. and Lundberg, A. 1987. Sexual selection in a monomorphic lek-breeding bird: correlates of male mating success in the great snipe *Gallinago media*. *Behavioural Ecology and Sociobiology* 21:211–216.

180 'In 1995 Höglund had been part of a Black Grouse study'. Rintamaki, P.T., Alatalo, R.V., Höglund, J. and Lundberg, A. 1995. Male territoriality and female choice on black grouse leks. *Animal Behaviour* 49:759–767.

181 'Old reports that the Pintail Snipe'. Byrkjedal, I. 1990. Song flight of the Pintail Snipe *Gallinago stenura* on the breeding grounds. *Ornis Scandinavica* 21:239–247.

182 'Buff-breasted Sandpipers in the North American Arctic'. Lanctot, R.B., Weatherhead, P.J., Kempenaers, B. and Scribner, K.T. 1998. Male traits, mating tactics and reproductive success in the buff-breasted sandpiper *Tryngites subruficollis*. *Animal Behaviour* 56:419–432. Lanctot, R., Scribner, K., Weatherhead, P.J. and Kempenaers, B. 1997. Lekking without a paradox in the Buff-Breasted Sandpiper. *American Naturalist* 149:1051–1070.

184 'In one study of Sharp-tailed Grouse'. Gratson, M.W. 1993. Sexual selection for increased male courtship and acoustic signals and against large male size at Sharp-tailed Grouse leks. *Evolution* 47:691–696. Lislevand, T., Figuerola, J. and Székely, T. 2009. Evolution of sexual size dimorphism in grouse and allies (Aves: Phasianidae) in relation to mating competition, fecundity demands and resource division. *Journal of Evolutionary Biology* 22:1895–1905.

8. Handicaps and Parasites

187 'From an Israeli scientist named Amotz Zahavi'. Zahavi, A. 1975. Mate selection – A selection for a handicap. *Journal of Theoretical Biology* 53:205–214.

190 'As Grafen put it'. Grafen, A. 1990. Biological signals as handicaps. *Journal of Theoretical Biology* 144: 517–546.

191 'Commented Penn and Számadó'. Penn, D.J. and Számadó, S. 2020. The Handicap Principle: how an erroneous hypothesis became a scientific principle. *Biological Reviews* 95:267–290.

191 'Grafen published a new study'. Biernaskie, J.M., Perry, J.C. and Grafen, A. 2018. A general model of biological signals, from cues to handicaps. *Evolution Letters* 2:201–209.

195 'As I chronicled in *The Red Queen*'. Ridley, M. 1993. *The Red Queen.* Penguin.

198 'Boyce treated half the male Sage Grouse'. Boyce, M. 1990. The Red Queen visits Sage Grouse leks. *American Zoologist* 30:263–270.

200 'Anders Møller, a Danish biologist based in Paris at the time'. Møller, A.P., Christe, P. and Lux, E. 1999. Parasitism, host immune function and sexual selection: a meta-analysis of parasite-mediated sexual selection. *Quarterly Review of Biology* 74:3–20. Møller, A.P., Dufva, R. and Erritzoe, J. 1998. Host immune function and sexual selection in birds. *Journal of Evolutionary Biology* 11:703–719.

203 'The Argentinian Lake Duck'. McCracken, K.G., Wilson, R.E., McCracken, P.J. and Johnson, K.P. 2001. Are ducks impressed by drakes' display? *Nature* 413:128–129.

203 'Brennan next dissected sixteen species of duck'. Birkhead, T.R. and Brennan, P. 2009. Elaborate vaginas and long phalli: Post-copulatory sexual selection in birds. *Biologist* 56:33–39.

206 'As Louise Mead and Stevan Arnold put it in 2004'. Mead, L.S. and Arnold, S.J. 2004. Quantitative genetic models of sexual selection. *Trends in Ecology and Evolution* 19:264–271.

208 'In 1997 Prum sent a paper to a high-profile journal'. Prum, R.O. 2010. The Lande–Kirkpatrick mechanism is the null model of evolution by intersexual selection: implications for meaning, honesty, and design in intersexual signals. *Evolution* 64:3085–3100.

212 'Prum's radical, ultra-Darwinian but anti-Wallacean view'. Prum, R.O. 2017. *The Evolution of Beauty.* Knopf Doubleday Publishing Group.

9. Paragon Peacock

217 'Shahla Yasmin and H.S.A. Yahya watched a population'. Yasmin, S. and Yahya, H. 1996. Correlates of mating success in Indian Peafowl. *Auk* 113:490–492.

218 'Back in 1984 we wrote up our results'. Rands, M.R., Ridley, M.W. and Lelliott, A.D. 1984. The social organization of feral Peafowl. *Animal Behaviour* 32:830–835.

219 'They then continuously watched a single Peacock lek'. Petrie, M. 2021. Evolution by sexual selection. *Frontiers in Ecology and Evolution* 9.

220 'Announced in the title of her paper, published in 1991'. Petrie, M., Halliday, T.R. and Sanders, C. 1991. Peahens prefer Peacocks with elaborate trains. *Animal Behaviour* 41:323–331.

221 'Jessica Yorzinski at the University of California Davis'. Yorzinski, J.L., Patricelli, G.L., Babcock, J.S., Pearson, J.M. and Platt, M.L. 2013. Through their eyes: selective attention in Peahens during courtship. *Journal of Experimental Biology* 216:3035–3046. Cyborg Peahen gives birds-eye view of Peacocks. Duke Today, 24 July 2013.

222 'In the winter of 1989 Petrie and Halliday did the obvious experiment'. Petrie, M., and Halliday, T. 1994. Experimental and natural changes in the Peacock's (*Pavo cristatus*) train can affect mating success. *Behavioural Ecology and Sociobiology* 35:213–217.

222 'Petrie then did an ingenious and careful experiment'. Petrie, M. 1994. Improved growth and survival of offspring of Peacocks with more elaborate trains. *Nature* 371:598–599.

223 'As Petrie pointed out, this was unexpected'. Pike, T.W., Cotgreave, P. and Petrie, M. 2009. Variation in the Peacock's train shows a genetic component. *Genetica* 135:7–11.

223 'In the early 2000s Adeline Loyau and her colleagues'. Loyau, A., Saint Jalme, M., Cagniant, C. et al. 2005. Multiple sexual advertisements honestly reflect health status in peacocks (*Pavo cristatus*). *Behavioural Ecology and Sociobiology* 58:552–557.

224 'The following year, Loyau performed another experiment'. Loyau, A., Gomez, D., Moureau, B., Théry, M., Hart, N.S., Saint Jalme, M., Bennett, A.T.D. and Sorci, G. 2007. Iridescent structurally based coloration of eyespots correlates with mating success in the peacock. *Behavioral Ecology* 18:1123–1131.

225 'Mariko Takahashi and her colleague Mariko Hiraiwa-Hasegawa'. Takahashi, M., Arita, H., Hiraiwa-Hasegawa, M. and Hasegawa, T. 2008. Peahens do not prefer peacocks with more elaborate trains. *Animal Behaviour* 75:1209–1219.

226 'Loyau and Petrie soon responded'. Loyau, A., Petrie, M., Saint Jalme, M. and Sorci, G. 2008. Do peahens not prefer peacocks with more elaborate trains? *Animal Behaviour* 76:e5–e9.

227 'Next it was the turn of Canadian Peacocks'. Dakin, R. and Montgomerie, R. 2011. Peahens prefer Peacocks displaying more eyespots, but rarely. *Animal Behaviour* 82:21–28.

227 'Marion Petrie teamed up with the Danish biologist Anders Møller'. Møller, A. and Petrie, M. 2002. Condition dependence, multiple sexual signals, and immunocompetence in peacocks. *Behavioral Ecology* 13:248–253.

228 'A total of forty-six males were penned separately'. Hale, M.L., Verduijn, M.H., Møller, A.P., Wolff, K. and Petrie, M. 2009. Is the peacock's train an honest signal of genetic quality at the major histocompatibility complex? *Journal of Evolutionary Biology* 22:1284–1294.

229 'In 2006, Marion Petrie raised another possibility altogether'. Petrie, M. and Roberts, G. 2007. Sexual selection and the evolution of evolvability. *Heredity* 98:198–205. Petrie, M. 2021. Evolution by sexual selection. *Frontiers in Ecology and Evolution* 9.

230 'In these open-minded days when even *Science* magazine'. Black, A. and Tylianakis, J.M. 2024. Teach Indigenous knowledge alongside science. *Science* 383:592–594.

232 'Maybe, I once argued'. Ridley, M. 1981. How the Peacock got its tail. *New Scientist* 91:398–401.

233 'Advanced by a Texas-based scientist named Michael Ryan'. Ryan, M. 1990. Sexual selection, sensory systems and sensory exploitation. *Oxford Surveys in Evolutionary Biology* 7:157–195.

235 'Green Peafowl seem to travel in small flocks'. Ponsena, P. 2022. Biological characteristics and breeding behaviours of Green Peafowl (*Pavo muticus* (*Linnaeus*)) in Huai Kha Khaeng Wildlife Sanctuary. *Thai Journal of Forestry* 7:303–313.

235 'Less polygyny in Green Peafowl than there is in Blue Peafowl'. Patil, A.B. and Vijay, N. 2022. Genome divergence across the Indo-Burman arc: a tale of two peacocks. *bioRxiv* 2022.05.27.493701.

235 'Blue Peafowl have more genes, and by a huge margin'. Chakraborty, A., Mondal, S., Mahajan, S. and Sharma, V.K. 2023. High-quality genome assemblies provide clues on the evolutionary advantage of blue peafowl over green peafowl. *bioRxiv* 2023.02.18.529039.

10. *The Riddle of the Ruff*

238 'There is an astonishing individuality about the Ruff'. Van Rhijn, J.G. 1991. *The Ruff*. T. and A.D. Poyser.

242 'Edmund Selous watched Ruff leks'. Selous, E. 1907. Observations tending to throw light on the question of sexual selection in birds, including a day-to-day diary on the breeding habits of the Ruff (*Machetes pugnax*). *Zoologist* 11:367–381.

244 'The independents often tolerate the satellites on their territories'. Tolliver, J.D.M., Kupan, K., Lank, D.B., Schindler, S. and Küpper, C. 2023. Fitness benefits from co-display favour subdominant male-male partnerships between phenotypes. *Animal Behaviour* 197: 131–154.

247 'In 2004 an observant Dutch farmer'. Ogden, L.E. 2014. In the world of ruffs, a male bird that's sneaky … and well endowed. Earth Touch News Network, 24 April 2014.

248 'DNA tests soon proved that Jukema was right'. Jukema, J. and Piersma, T. 2006. Permanent female mimics in a lekking shorebird. *Biology Letters* 2:161–164.

249 'Both teams succeeded in simultaneously mapping the genome of the Ruff'. Küpper, C., Stocks, M., Risse, J.E. et al. 2016. A supergene determines highly divergent male reproductive morphs in the Ruff. *Nature Genetics* 48:79–83. Lamichhaney, S., Fan, G., Widemo, F. et al. 2016. Structural genomic changes underlie alternative reproductive strategies in the Ruff (*Philomachus pugnax*). *Nature Genetics* 48:84–88.

249 'Much more recently, around seventy thousand years ago'. Hill, J., Enbody, E.D., Bi, H., Lamichhaney, S., Lei, W., Chen, J., Wei, C., Liu, Y., Schwochow, D., Younis, S., Widemo, F. and Andersson, L. 2023. Low mutation load in a supergene underpinning alternative male mating strategies in Ruff (*Calidris pugnax*). *Molecular Biology and Evolution* 40: msad224.

250 'A large number of deletions, insertions and duplications'. Maney, D.L. and Küpper, C. 2022. Supergenes on steroids. *Philosophical Transactions of the Royal Society* B 377:20200507.

252 'The rock, paper, scissors game'. Sinervo, B. and Lively, C. 1996. The rock–paper–scissors game and the evolution of alternative male strategies. *Nature* 380:240–243.

253 'The question of how the inversion supergene survives'. Baguette, M., Bataille, B. and Stevens, V.M. 2022. Evolutionary ecology of fixed alternative male mating strategies in the Ruff (*Calidris pugnax*). *Diversity* 14:307.

254 'Clemens Küpper and his colleague Lina Giraldo-Deck'. Giraldo-Deck, L.M., Loveland, J.L., Goymann, W., Tschirren, B., Burke, T., Kempenaers, B., Lank, D.B. and Küpper, C. 2022. Intralocus conflicts associated with a supergene. *Nature Communications* 13:1384.

255 'Chimps' testes are more than four times the size of Gorillas''. Harcourt, A., Harvey, P., Larson, S. et al. 1981. Testis weight, body weight and breeding system in primates. *Nature* 293:55–57.

256 'They are indeed polyandrous'. Lank, D.B. et al. 2002. High frequency of polyandry in a lek mating system. *Behavioral Ecology* 13:209–215.

11. An Aesthetic Sense

260 'Alfred Russel Wallace's account'. Wallace, A.R. 1869. *The Malay Archipelago*. Macmillan.

261 'Jan Huyghen van Linschoten wrote in 1596'. Van Linschoten, J.H. 1596. Itinerario: Voyage ofte schipvaert van Jan Huygen van Linschoten naer Oost ofte Portugaels Indien, 1579–1592. (Published in English 1598.)

261 'The indefatigable Tim Laman'. Scholes, E. and Laman, T. 2012. The quest to film and photograph every species of Bird-of-Paradise. Allaboutbirds.org, Cornell University.

263 'San Diego Zoo has tried for years to breed Raggianas'. Theule, J. and Rimlinger, D. 2022. Breeding history and husbandry of the Raggiana bird-of-paradise (*Paradisaea raggiana*). *Zoo Biology* 42:162–170.

266 'Can absorb up to a hundred times as much light as those on a crow'. McCoy, D.E., Feo, T., Harvey, T.A. and Prum, R.O. 2018. Structural absorption by barbule microstructures of super black bird of paradise feathers. *Nature Communications* 9:1.

267 'The Vogelkop Bird of Paradise'. Scholes, E. and Laman, T.G. 2018. Distinctive courtship phenotype of the Vogelkop Superb Bird-of-Paradise *Lophorina niedda* Mayr, 1930 confirms new species status. *PeerJ* 6:e4621.

267 'The physical properties of the feathers on Lawes' Parotia'. Wilts, B.D., Michielson, K., De Raedt, H. and Stavenga, D.K. 2014. Sparkling feather reflections of a bird-of-paradise explained by finite-difference time-domain modelling. *Proceedings of the National Academy of Sciences* 111:4363–4368.

268 'This is an unexpected finding'. Ligon, R.A., Diaz, C.D., Morano, J.L., Troscianko, J., Stevens, M., Moskeland, A. et al. 2018. Evolution of correlated complexity in the radically different courtship signals of birds-of-paradise. *PLoS Biology* 16: e2006962.

275 'An analysis of the brains of bowerbirds'. Day, L.B., Westcott, D.A. and Oster, D.H. 2005. Evolution of bower complexity and cerebellum size in bowerbirds. *Brain, Behavior and Evolution* 66:62–72.

277 'Endler did various experiments'. Endler, J.A., Endler, L.C. and Doerr, N.R. 2010. Great Bowerbirds create theaters with forced perspective when seen by their audience. *Current Biology* 20:1679–1684.

278 'Clifford and Dawn Frith began studying the birds in exhaustive detail'. Frith, C. and Frith, D. 2008. *Bowerbirds: nature, art, history.* Frith and Frith.

279 'A neat, if slightly unkind, experiment'. Coleman, S., Patricelli, G. and Borgia, G. 2004. Variable female preferences drive complex male displays. *Nature* 428:742–745.

12. How Mate Choice Shaped the Human Mind

285 'The same appears to be true of human beings'. Shackelford, T., Goetz, A., LaMunyon, C., Pham, M. and Pound, N. 2015. Human sperm competition. 10.1002/9781119125563.evpsych115.

286 'Polygamy became almost the norm'. Ross, C.T., Borgerhoff Mulder, M. et al. 2018. Greater wealth inequality, less polygyny: rethinking the polygyny threshold model. *Journal of the Royal Society Interface* 1520180035.

288 'The Cambridge zoologist John Madden'. Madden, J.R. 2008. Do bowerbirds exhibit cultures? *Animal Cognition* 11:1–12.

288 'Nature operates via nurture, not versus it'. Ridley, M. 2003. *Nature via Nurture.* HarperCollins.

288 'As the anthropologist Joe Henrich has documented'. Henrich, J. 2016. *The Secret of Our Success: how culture is driving human evolution, domesticating our species, and making us smarter.* Princeton University Press.

290	'Raymond Corbey of Leiden University'. Corbey, R., Jagich, A., Vaesen, K. and Collard, M. 2016. The Acheulean Handaxe: more like a bird's song than a Beatles' tune? *Evolutionary Anthropology* 25:6–19.

291	'Preferred cultural conservatism'. Finkel, M. and Barkai, R. 2018. The Acheulean Handaxe technological persistence: a case of preferred cultural conservatism? *Proceedings of the Prehistoric Society* 2018:1–19.

291	'The American psychologist James Mark Baldwin'. Baldwin, J.M. 1896. A new factor in evolution. *American Naturalist* 30:441–451.

292	'As Corbey puts it'. Corbey, R. 2020. Baldwin effects in early stone tools. *Evolutionary Anthropology* 29:237–244.

293	'Marek Kohn and Steve Mithen'. Kohn, M. and Mithen, S. 1999. Handaxes: products of sexual selection? *Antiquity* 73:518–526.

294	'At Sunghir, north-east of Moscow'. Trinkaus, E. and Buzhilova, A.P. 2018. Diversity and differential disposal of the dead at Sunghir. *Antiquity* 92:7–21.

294	'In Jordan, a Neolithic child'. Alarashi, H. et al. 2023. Threads of memory: reviving the ornament of a dead child at the Neolithic village of Ba`ja (Jordan). *PLoS One.* https://doi.org/10.1371/journal.pone.0288075.

294	'In Iraq, a Bronze Age queen of Ur'. Durn, S. 2022. The most lavish Mesopotamian tomb ever found belongs to a woman. *Atlas Obscura*, 10 February 2022.

295	'As Ernst Grosse wrote in an 1897 book'. Grosse, E. 1897. *The Beginnings of Art.* D. Appleton.

295	'Sexual selection cuts this Gordian knot'. Miller, G.F. 2016. Art-making evolved mostly to attract mates. In *On the Origins of Art* [exhibition catalog] pp. 163–213. Hobart, Tasmania: Museum of Old and New Art.

296	'Helen Clegg, Daniel Nettle and Dorothy Miell'. Clegg, H., Miell, D. and Nettle, D. 2011. Status and mating success amongst visual artists. *Frontiers in Psychology* 2:310.

296	'In one survey by Helen Fisher of a thousand people'. Fisher, H. [no date] Humor … A Hidden Aphrodisiac. Helenfisher.com

297	'As Miller put it in his book *The Mating Mind*'. Miller, G. 2000. *The Mating Mind.* William Heinemann.

Epilogue: Saving the Black Grouse

305	'This bird could go extinct in Britain'. Warren, P. 2022. Black Grouse under pressure. Game and Wildlife Conservation Trust.

305 'Gilbert White, the naturalist vicar of Selborne'. White, G. 1789. *The Natural History of Selborne*.

308 'A sharp decline in desert tortoise numbers in a California desert'. Boarman, W. 2003. Managing a subsidized predator population: reducing common raven predation on desert tortoises. *Environmental Management* 32:205–217.

308 'Generalist "meso-predators"'. Prugh, L.R., Stoner, C.J., Epps, C.W., Bean, W.T., Ripple, W.J., Laliberte, A.S. and Brashares, J.S. 2009. The rise of the mesopredator. *BioScience* 59:779–791.

308 'The farmer and writer Patrick Laurie'. Laurie, P. 2012. *The Black Grouse*. Merlin Unwin Books.

309 'Hen Harriers for example'. Further increase in English Hen Harrier numbers recorded in 2023. Naturalengland.blog.gov.uk, 16 September 2023.

310 'Open, treeless moorland is also much more natural'. Gallego-Sala, A.V., Charman, D.J., Harrison, S.P., Li, G. and Prentice, I.C. 2015. Climate-driven expansion of blanket bogs in Britain during the Holocene. *Climate of the Past Discussions* 11:4811–4832. Fenton, J. 2024. *Landscape Change in the Scottish Highlands*. Whittles Publishing.

Index